The New Chimpanzee

CRAIG STANFORD

The New Chimpanzee

A Twenty-First-Century Portrait
of Our Closest Kin

Harvard University Press

Cambridge, Massachusetts, and London, England   2018

Copyright © 2018 by the President and Fellows of Harvard College
All rights reserved
Printed in the United States of America
First printing

Library of Congress Cataloging-in-Publication Data

Names: Stanford, Craig B. (Craig Britton), 1956– author.
  Title: The new chimpanzee : a twenty-first-century portrait of our closest kin / Craig Stanford.
Description: Cambridge, Massachusetts : Harvard University Press, 2018. | Includes bibliographical references and index.
Identifiers: LCCN 2017040032 | ISBN 9780674977112 (hardcover : alk. paper)
Subjects: LCSH: Chimpanzees—Behavior. | Primatology.
Classification: LCC QL737.P94 S727 2018 | DDC 599.885—dc23
  LC record available at https://lccn.loc.gov/2017040032

Dedicated to the generations of chimpanzee researchers from Jane Goodall to the present day

Contents

Preface   *ix*

1 Watching Chimpanzees   *1*
2 Fission, Fusion, and Food   *19*
3 Politics Is War without Bloodshed   *41*
4 War for Peace   *66*
5 Sex and Reproduction   *91*
6 Growing Up Chimpanzee   *114*
7 Why Chimpanzees Hunt   *130*
8 Got Culture?   *154*
9 Blood Is Thicker   *176*
10 Ape into Human   *191*

Notes   *209*
Bibliography   *223*
Acknowledgments   *259*
Credits   *263*
Index   *265*

Preface

Over the past two decades, scientists have made dramatic discoveries about chimpanzees that will change the way we understand both human nature and the apes themselves. Although there is a rich history of chimpanzee field research going back nearly sixty years, almost all the findings discussed in this book have been made just since the turn of the millennium. From genomics to cultural traditions, we'll consider our close kin in a new light and ask what this information may mean for a new and improved understanding of human nature.

Studying wild chimpanzees is the profession of a very small number of people in the world. At any one time there are probably fewer than a hundred scientists and their students actively engaged in chimpanzee field observation and study. The number of full-time professors in American universities whose careers are focused mainly on wild chimpanzee research is perhaps a dozen. Add in the scholars and conservationists doing work in related areas, and the global army of chimpanzee watchers is a few hundred strong. The available funding for the work they do is a fraction of that given to scientists in other endeavors. Yet the results of new studies are front-page news and are rightly touted in the international media for the clues they provide about human nature.

My own involvement with chimpanzees came about fortuitously. In the late 1980s I was conducting my doctoral research in Bangladesh on a previously little-known monkey called the capped langur. I was living in a ramshackle cabin on stilts at the edge of a rice paddy, spending my days following a group of the monkeys on their daily rounds in the nearby forest. Capped langurs are handsome animals, their gray backs set off by a flame-orange coat underneath and a black mask of skin for a face. Unfortunately, their behavior is not as interesting; they traveled only a hundred meters per day and spent nearly all their waking hours calmly munching on foliage. The most interesting observation I made in thousands of hours with the langurs was the fatal attack on an old female by a pack of jackals. Jackals were not thought to prey on animals as large as eight-kilogram monkeys, but with the extirpation of leopards and tigers in the area, they may have taken on that role. As I strolled along behind the group one afternoon, observing the matriarch feeding on the ground right in front of me, a pair of jackals burst from a thicket, grabbed her, and dragged her off. It was a vivid demonstration for me of the potential for predators to make a powerful impact on the survival of an individual, and on the population of monkeys in this forest.

As I looked ahead to the completion of my PhD and considered postdoctoral options, I sent letters to a number of primate researchers in Africa and Asia proposing projects that involved the study of predation's effects on wild primate populations. A colleague suggested I write to Jane Goodall. Goodall's field site in Tanzania had been attacked by a rebel militia in 1975. Four Western students were kidnapped and held for ransom in neighboring Zaire (now the Democratic Republic of the Congo). Although all were eventually released unharmed, the park had been generally off-limits to visiting researchers for more than a decade. I mailed a thin blue aerogram—this was pre-Internet—expecting no reply. When I returned to Berkeley months later, a letter from Goodall was waiting, inviting me to come to Gombe to study the predator-prey interactions between chimpanzees and the red colobus monkeys, whose flesh they so relish. A year later, with a permit from the Tanzanian government and a shoestring budget in hand, I arrived to begin several years of back-and-forth travel to Gombe to study the hunting behavior of chimpanzees and its impact on the behavior and population biology of the monkeys they hunt.

The world's most famous study of animal behavior is located in a former British colonial hunting reserve, now a tiny but important jewel in Tanzania's national park system. It's an oblong strip of forest and hills, about ten miles long and two miles wide, hugging the shore of Lake Tanganyika, two hours by boat from the harbor town of Kigoma. Before Goodall's arrival, Kigoma was best known for its harbor and its proximity to Ujiji, where the newspaper reporter Henry Morton Stanley found the missionary doctor and explorer David Livingston in 1871. It was a sleepy port town with one main dirt road, a few cafés, and a lot of ramshackle market stalls. These days Kigoma is bustling; it's the jumping-off point for ecotourists headed to either Gombe or Mahale National Parks to see wild chimpanzees. Beginning in the 1960s it was the home of Goodall and a team of Tanzanian assistants, soon joined by students from North America and Europe who documented the intimate details of the lives of wild apes. It is sacred ground for any student of animal behavior.

I had come to Gombe mainly to study red colobus monkeys—the favored prey animal when chimpanzees go hunting—and their relationship with their predators. Unlike most researchers who arrive at Gombe to study chimpanzees, I was fairly ignorant about their celebrity status. I knew that each member of the same matriline bore a name starting with the same letter, but the names Fifi, Frodo, Gremlin, and Goblin bore no special meaning to me at the start of my study. This would soon change; spending long hours in close quarters with chimpanzees, you cannot help becoming immersed in their lives and personalities. The daily life of a researcher is organized around the daily lives of the animals; you go where they go, when they go, resting when they rest and sweating up steep hills right behind them.

This book is about the lives of chimpanzees living where they belong, in the tropical forests of Africa. Most of us who have spent time with wild chimpanzees have an uneasy relationship with captive research and the ethics of keeping and studying apes in zoos, laboratories, and the like. Psychologists working in laboratories and primate centers have made amazing discoveries about the workings of the chimpanzee mind. These discoveries hold great promise for a deeper understanding of our own intellect and the meaning of intelligence. On the other hand, these researchers work with chimpanzees that are locked up in enclosures that, however artfully designed, cannot mask the fact that the apes are prisoners. In the best-case scenario, large and well-funded primate centers maintain their chimpanzees in large social groups that occupy spacious outdoor enclosures. In such places, detailed observations of social behavior are possible that could never be accomplished in the rugged, dense forests in which chimpanzees naturally live. But spaciousness is relative. A very lucky captive chimpanzee might spend his life on a well-landscaped two-acre island. The same ape, if raised in Africa, would spend a lifetime traversing up to fifty square kilometers of forest. Captivity provides freedom from hunger, predators, and disease, but with the loss of environmental enrichment beyond measure. A cage is a cage, no matter how large or well designed. Moreover, research on behaviors that evolved in an African forest should obviously be done in an African forest. For this reason I have chosen to focus on what we have recently learned about chimpanzees in the wild, and only occasionally introduce captive findings, exciting though they may be in their own right.

The book is organized as a series of narratives about each of the major areas of recent research. Chapter 1, "Watching Chimpanzees," sets the stage by describing the close evolutionary relationship between humans and their ape relatives and discussing many of the issues current in chimpanzee field research. Chapter 2, "Fission, Fusion, and Food," is about the complicated nature of chimpanzee society and the role that both ovulating females and food availability play in it. In Chapter 3, "Politics Is War without Bloodshed," I consider the Machiavellian and fascinating world of chimpanzee politics. Chapter 4, "War for Peace," focuses on the nature and causes of violence within and between chimpanzee communities. Chapter 5, "Sex and Reproduction," is about the rather complicated sexual politics of chimpanzee life. Chapter 6, "Growing Up Chimpanzee," describes the early lives of male and female chimpanzees and the factors that turn them into successful or not-so-successful adults. Chapter 7, "Why Chimpanzees Hunt," describes chimpanzee meat-eating behavior, which shocked the world when first reported by Goodall in the early 1960s. I reflect on my own work on chimpanzee hunting and recount the ensuing debate over the meaning of meat in chimpanzee society.

Chimpanzees are the most technologically gifted creatures on the planet besides ourselves. In Chapter 8, "Got Culture?," I examine chimpanzee intelligence as evidenced by the many recent discoveries of tool use among wild chimpanzees. Chapter 9, "Blood Is Thicker," is about the emerging field of ape genomics. I share what we have learned in the past few years about the genetics of chimpanzees and what that may tell us about our own evolutionary history. Chapter 10, "Ape into Human," concludes the book with some lessons that we can learn from chimpanzees in order to better understand ourselves. It is also about the use and misuse of extrapolations from ape behavior to that of early humans.

My hope is that readers will appreciate chimpanzees for what they are—not underevolved humans or caricatures of ourselves, but perhaps the most interesting of all the species of nonhuman animals with which we share our planet. The gift of the chimpanzee is the vista we are offered of ourselves. It is a gift that is in danger of disappearing as we destroy the chimpanzees' natural world and drive them toward extinction. A tiny fraction of chimpanzees live in protected sanctuaries, where their health is monitored and we are aware of every problem that faces them. The other 150,000 chimps living in the forests of Africa are still unknown, unmonitored, and in dire need of protection. To gain a fuller sense of what we will lose if chimpanzees cease to exist in the wild, read on.

CHAPTER 1

Watching Chimpanzees

THE RECORDED HISTORY of humanity dates back to the Sumerians and ancient Egyptians in the fourth millennium B.C. Human prehistory extends back three hundred thousand years further to the emergence of our species, *Homo sapiens,* from far more primitive kinds of hominins. But although chimpanzees have an evolutionary history that goes back at least a million years, their recorded history begins only with the words of Jane Goodall, writing in 1960, barely a half century ago. Imagine if chimpanzee history had been recorded for the previous thousand generations. We would learn of powerful alpha males and females, epic wars, diverse cultures, and heart-stopping hunting raids. But there are no ape historians, so we are left with only the present.

Fortunately, there are many ongoing, long-term studies of chimpanzees, both in the forests of Africa where they evolved and in captivity. The history of chimpanzee research continues today through the work of a small army of primatologists. We have studied their behavior, diet, anatomy, and psychology for many decades, and in recent years the rise of molecular genetic technology has allowed us to address new questions about their close kinship with us.

People have been watching chimpanzees for thousands of years. The earliest apelike humans may have even had a competitive relationship with their ape cousins when both lived in or near the same African forests. Certainly by the time *Homo sapiens* evolved around three hundred thousand years ago, they would have pondered the hairy apes living in forests near their camps. People in rural Africa today hold a wide range of attitudes about living with chimpanzees as neighbors. In some areas they hunt and eat them. They regard them as food competitors when the apes raid their crop fields. They live in fear of them too, knowing full well how dangerous a male chimpanzee can be. There are a number of records of wild chimpanzees coming into a village to attack and drag off a screaming child. On the other hand, people keep baby chimpanzees as pets. All the while, they have marveled at the intelligence and humanness of their ape neighbors.

Observing Wild Chimpanzees

There is a Zen quality to the craft of studying wild animals—long and tedious hours of quiet observation in hopes of getting a glimpse of behavior that no one has seen before. If you're not good at living inside your own head much of the time, you won't be able to cope with it. Even on the best days, chimpanzees spend a lot of time doing very little, which leaves you waiting for something to happen while staring at hairy black backs sitting in dense foliage. It sounds glamorous to say you're spending your work hours in an African forest watching great apes rather than at an office desk in Los Angeles, but much of the work of watching any wild animal happens at a glacial pace. In between brief episodes of sex, violence, and soap-opera drama in chimpanzee society, there is the business of eating, which they spend hours each day doing. They also spend a few hours during the day napping or relaxing, plus the twelve or so hours of nighttime during which they slumber in their nests. So a large percentage of one's time is spent watching the animals snoozing, staring blankly into space, or munching contentedly on fruit.

Doing very little is part of the rhythm of daily life for all of us, and I suppose it holds its own fascination, but this is the equivalent of a cultural anthropologist's watching the people she studies while they sleep on the couch. I have colleagues who study big cats on the African plains, which sounds adventurous and glamorous. But in between the occasional hunts, the scientists sit behind the wheel of a Toyota pickup, wilting in the sun. By comparison, hours spent sitting in a leafy glade waiting for the chimpanzees to wake up and do something is not too bad. In any animal-watching study, the animals set the pace and dictate your daily habits. They also dictate your emotional state. When I was studying chimpanzee hunting behavior in Gombe Stream National Park, Tanzania, my research involved recording a behavior that happened only once or twice a week at most. Sometimes two or three weeks would pass without any hunting; since my task was to observe and understand that behavior, it left me feeling that I wasn't accomplishing anything (because I wasn't). There is always something happening in the forest, from chance observations of rare forms of tool use by the chimps, to a rare bird or python sighting. But when you're in the field to do research, it's not fun to end each day for weeks on end without any new data.

The process of initiating a study of wild chimpanzees is different from that for most wild animals. We spend years trying to accustom them to our approach so we can see the intimate details of their lives without interfering with their natural behavior. Wild chimpanzees have a healthy fear of people; by making visual contact with them every day and showing that you're utterly harmless, you gradually acclimate them to your presence. Each month the contact distance is shortened, and eventually you can sit in the middle of a group of chimpanzees and be accepted as an unobtrusive part of the landscape as they groom, mate, or fight. The first morning that I went out at Gombe to follow chimps, we arrived at their night nests in predawn darkness. As I picked my way across a steep slope under their nesting trees, I placed my hand on a boulder to steady myself. The boulder was the shoulder of an adult male chimpanzee, groggy and half-awake. We both flinched, I muttered an involuntary "excuse me," and I got my first sense of just how utterly accustomed to humans a wild ape can become. Such approachability is the key to immersive research.

It's not difficult to observe large, social wild animals that are active in the daytime and live in open country, like giraffes or elephants. The challenge is getting close enough to see the inner details of their lives. Chimpanzees, on the other hand, live in dense forests, spend a lot of time high in trees, and are shy wherever they have a history of being harassed or hunted by people. The hard work of getting them to tolerate your presence is rewarded by the incredible privilege of sharing the hours of their lives with them, bearing witness to their triumphs and tragedies.

Watching can involve high drama. For example, you could be sitting in the middle of a cluster of napping chimpanzees in a thicket when chaos suddenly breaks out, or watching a hunt unfolding in the trees overhead as a terrified monkey falls to the ground at your feet and is leaped on by the predatory chimps. This is what I meant earlier by the episodic nature of observation. After hours of mind-numbing inactivity, all hell may break loose and key events in the animals' lives can happen in the space of a few seconds. Without a video recording of the action, and with the chimpanzees moving in and out of dense foliage, reconstructing the action later is usually impossible. For perspective, however, I think of all my colleagues who study animals in environments far more impenetrable than a forest: Dolphins are fascinating, but they live their lives in the murky depths of the ocean. Many birds fly ten kilometers every morning, well beyond the ability of a researcher to keep up. Even orangutans, with their treetop habitat and solitary nature, are a test of a primatologist's persistence and patience. Compared to the observers of many other animals, chimpanzee watchers don't have a bad deal.

For generations, cultural anthropologists have inserted themselves in the daily lives of their study subjects. They do so in order to be participant observers. They sit around a campfire with villagers and chat about their lives. As primatologists, we seek the opposite—to be up close without being personal. We try to be present without being present. This can be a difficult task. It means we don't eat during an entire day in the forest for fear that the chimpanzees will learn to associate researchers with a free meal. It means maintaining a certain proximity but not coming too close for fear we might unwittingly pass a cold virus or flu to them or in some way influence their next move. The history of chimpanzee research is pockmarked by instances in which researchers unintentionally interfered with the lives of the animals they were trying to study.

Modern field studies began with Goodall, and so did the practice of manipulating the apes' behavior in order to make them more watchable. The early days of Goodall's work were full of frustration. She listened to distant chimpanzees and caught fleeting glimpses of them from her hilltop observation post. Eventually a white-chinned old male chimp she called David Greybeard approached her little tented camp on the beach of Lake Tanganyika to feed in a nearby palm tree. He became the unofficial ambassador who bridged the gap between the young primatologist and the animals she had come to Gombe to study. David Greybeard led others to the same spot, and on one of the visits he took bananas from Goodall's camp. Goodall began to put out bunches of bananas around her camp, and later she cleared a patch of forest to create a feeding station in hopes of drawing the wary apes closer so that she could film them. It worked all too well. Soon most of the local chimpanzees were emerging from the forest shadows and allowing themselves to be observed.[1]

Provisioning the Gombe chimpanzees with bananas fast-forwarded the process of acclimation by years. In later studies elsewhere, acclimating the apes to the presence of researchers would take far longer. In the earliest days of primate study, there was little understanding of how the acclimation process—what primatologists call habituation—could alter the daily lives of the animals we had come to study. Putting a pile of fruit in a clearing creates, in effect, the biggest fruit tree in the forest, and the chimpanzees come regularly to that spot to reap the human-provided bounty. This leads to an unnatural concentration of chimpanzees hanging around the same place for hours at a time, with increased fighting as banana-crazed apes squabble over their prizes. It may even exaggerate natural dominance relationships among them. But in Goodall's earliest days, our understanding of chimpanzee behavior and how best to study it was in its infancy. The bananas enabled her to watch chimpanzees in closer proximity than had ever been dreamed of before, and the results were dramatic. After months of frustration spent trying to stealthily approach the animals in the forest, Goodall could sit and watch them from a few feet away.

What does it mean to watch a wild animal's life unfold? Most laboratory scientists intentionally manipulate circumstances to study the outcome: they experiment. Some experimental scientists do their work in the natural world but still try to modify some aspect of their study in order to see the effects. In a few branches of science, the researcher gathers information without manipulating anything. Paleontologists, for example, search for new fossils that they can use to test ideas generated by older fossil discoveries. But it's very rare for a scientist today to simply watch and record his or her observations. In an earlier era, this was called natural history, and it is today held in fairly low regard by a new generation of scientists who apply cutting-edge tools to their work. I have colleagues who collect urine from wild apes to study the hormonal profiles of males and females and how they are influenced by the environment. Others use loudspeakers to play back calls of other chimpanzees in order to record how their study subjects respond to the perceived intruders in their territory. Still others collect DNA from feces to study the role that kinship plays in social relationships. All of these are exciting new techniques that lend themselves to natural experiments. But the fundamental part of all these studies is still observation of the animals.

We normally don't intervene in the chimpanzees' lives. We don't feed them when they're starving during a famine or doctor them when they're sick. We also try to avoid becoming the object of their ire. Some male chimpanzees have learned how much fun it is to bully human researchers. Sometimes they slap us as they charge past while displaying their status to other chimps. Occasionally they become outright aggressive and anyone standing nearby must take great care. One Gombe chimp in particular—a large male named Frodo—developed an aggressive habit of bullying researchers. He once knocked me down and sat down beside me, first tapping my head and then grooming me briefly before deciding I was too boring an object to merit more of his attention.

There have been times when researchers have intervened to save wild chimpanzees in danger. Goblin, a male in Gombe, was an impressive adult of some fame within chimpanzee circles for having achieved alpha status at a very young age and having held it for many years. After losing his top rank, he made a comeback attempt and was badly injured by other males, who were less than impressed by his bravado. He lay dying on the forest floor for days, his wounds festering. A decision was made that researchers should intervene to attempt to save his life, partly in order to see how the remarkable life he had led up to that point would play out in the longer term. Researchers brought water and antibiotic-laced bananas to the thicket in which he lay, and at one point they even disinfected his open wounds. Goblin recovered and went on to live many more years. His life had already been altered by the research project established around him; the decision to try to save him rather than let him die seemed obvious. Was the intervention ethical? It depends on whether we place more value on an individual's life or on avoiding disruption of the natural progression of leadership, life, and death in the community.

When Goodall submitted her earliest scientific papers for publication, reviewers were aghast that she regarded her chimpanzees as unique individuals. Goodall pushed back, pointing out that apes possess personas that play key roles in their social lives within a community. Fifty years later, no one would dispute her opinion, but it took time to change the minds of those in the scientific community who were more accustomed to studies of rats and mice than apes. That perspective has undergone a full 180-degree turn away from the earlier perspective that the primatologist Frans de Waal of Emory University has called "anthropodenial."[2] We now recognize that chimpanzees have individual personas and the same basic set of emotional and psychological underpinnings that humans do.

One of the greatest pitfalls of chimpanzee research today is nevertheless anthropomorphism, and it begins with researchers watching and interpreting the actions of their subjects. If you think your dog displays emotions such as shame, embarrassment, and guilt—most people do—then you'd be far more easily convinced that chimpanzees do too. Zoogoers know it too, although they may express it with embarrassed giggles at the sight of ape sex and sanitation. We have been applying human traits to great apes for centuries, and while children and casual observers may be the worst offenders, scientists are by no means immune. The reason is obvious; we know too well that the roots of their nature and of our own are one and the same. While dismissing the inner emotional and psychological life of animals as simply responses to stimuli is naïve, so is assuming there is always a connection between our mental states and those of other primates.

We would love to know more about the ancestral roots of the behavior of chimpanzees. There is virtually no fossil record for the direct ancestry of modern chimpanzees, save a few teeth from a chimpanzee-like fossil ape that lived in East Africa about a half million years ago. We think the lack of fossils of modern apes is due to the wet forest habitat of chimpanzees across much of central Africa, where dead apes decompose long before they can be turned into fossils, and where fossil hunters therefore rarely go in search of them. If Africa is the cradle of humankind and of human nature, it is equally the birthplace of our primate kin. Modern chimpanzees arose from a diverse and widespread array of ancient apes that inhabited the forests of Africa starting more than twenty million years ago. The four modern African apes that remain—the chimpanzee, the bonobo, and the lowland and mountain species of gorilla—are the remnants of a once-great radiation. The chimpanzee is by far the most environmentally versatile of these, inhabiting the dry woodlands of Tanzania in the far east of Africa and the savannah woodlands of Senegal some four thousand kilometers to the west, covering more than two million square kilometers. In between lies their true stronghold, the vast forests of the Congo basin. Across this expanse, perhaps two hundred thousand chimpanzees remain.[3]

The History of Chimp Watching

African people have been watching chimpanzees for thousands of years in the forests near their village homes, and scattered reports by early European explorers mention humanlike primates in forests there as well. The earliest real studies of modern chimpanzees were of their bodies, not their behavior. The carcasses of wild chimpanzees shot by hunters sometimes made their way to museums in Europe. In the late seventeenth century, British anatomist Edward Tyson received a chimpanzee from Angola, in southwestern Africa, that had died en route to London and been packed in salt for preservation. Tyson performed a dissection and made the first comparisons of form and function between human and ape. It did not take much expertise to see the striking similarities.[4] Captive chimpanzee studies began a few decades later, but for hundreds of years most attention was focused on the cognitive abilities of the apes in captivity. Charles Darwin was fascinated by an orangutan in the London Zoo.

The earliest field studies to be published were by trophy hunters and the earliest generation of wildlife photographers. By the early twentieth century, interest in apes had intensified enough that safari hunters had become naturalists in hopes of catching glimpses of their quarry alive, at least briefly, before shooting them. Soon after the invention of the movie camera, the first wildlife safaris were launched in search of apes. In 1909, Carl Akeley—the greatest taxidermist of his day—accompanied President Teddy Roosevelt on one of his African big game safaris. He later returned to Africa on his own expedition in search of gorillas to stuff for the new African wing of the American Museum of Natural History in New York. But after killing several mountain gorillas in the Virunga volcanoes, he felt remorse at the murder of the creatures. His observations of gorilla behavior were brief, but they stood for decades as the most detailed scientific study of the animals in the wild.[5] Beyond collecting expeditions that sometimes involved observations of the quarry before killing them, very little was yet known about the behavior of any wild ape.

In the Shadow of Jane

Our utter ignorance about wild great apes changed in the late 1950s. Legendary wildlife biologist George Schaller headed to east-central Africa in 1959 to conduct the first systematic field study of gorillas. Equally legendary Japanese primatologists Junichiro Itani and Kinji Imanishi had surveyed gorillas in eastern Africa, concluding their work at precisely the same time as Schaller. A groundswell was in the making. Itani later corresponded with famed fossil hunter Louis Leakey about the possibility of studying wild chimpanzees in a place called Gombe in westernmost Tanzania. But in July of that same year, Jane Goodall had made camp on the gravelly shore of Lake Tanganyika, sent by Leakey to carry out the study that he had been yearning for someone to do. Itani actually visited Goodall in Gombe that year, hoping to join in the research himself. Seeing that she appeared to have established her camp at Gombe for the long haul—Leakey had allegedly suggested to Itani that Goodall's study might be of short duration—Itani and his colleagues moved a hundred kilometers down the lakeshore and established their own camp in the Mahale Mountains. In the United States, biological anthropologist Sherwood Washburn was inspiring a new generation of graduate students to conduct field studies of wild monkeys and apes. About this time, Dutch scientist Adrian Kortlandt went to the Belgian Congo in search of chimpanzees and found them in forests bordering banana plantations where they raided crops. Kortlandt built an observation platform eighty feet up in a large tree and spent many hours watching the apes go about their lives. He also conducted the first primate field experiment, placing a stuffed leopard in the chimpanzees' territory and observing their panicked reaction.[6]

Goodall was the pioneer, but she was by no means the only researcher interested in "real" chimpanzees in the early 1960s. However, it was Goodall who stayed on and made the hallmark discoveries that changed our views of primate behavior and human nature. She made the historic discoveries of tool use and hunting, and she later provided key information on chimpanzee society that came with multigenerational research. It's hard to convey to students today the impact of Goodall's discoveries in the 1960s and 1970s. At the time, a long-term field study was measured in weeks or months, not years or decades. The idea that chimpanzees might be skilled toolmakers, cooperative hunters, or occasionally bloodthirsty killers, or that they might show the clear hallmarks of a simple humanlike culture, sounded like science fiction in 1960. By the 1970s it was textbook chimpanzee behavior.

Had Goodall followed a more conventional path, she might not have become the pioneering scientist and household name she is today. Instead of attending university and then graduate school, followed by a stint of six months or a year in Africa (which would have been lengthy at that time for a career-minded young primatologist), Goodall fulfilled a girlhood dream by traveling to East Africa with a friend in 1957. She met Leakey, who offered her a clerical job in the Coryndon Museum in Nairobi (now part of the National Museums of Kenya). He was looking for an eager young person to go out to Gombe Stream Reserve to watch wild chimpanzees. Gombe had been a hunting reserve during British colonial days. By 1960 most of its big animals were gone, but it was still a fairly wild strip of forest along Lake Tanganyika with reports of a population of chimpanzees. For decades, Leakey had been searching for human fossils in arid East African landscapes that had been forested lakeshores three million years earlier. He passionately wanted to understand the lives of the earliest humans, and he knew that the lives of wild chimpanzees provide the best possible window. Leakey had considered other candidates, but in the end it was Goodall who landed on the shore of the forest reserve to set up a camp and attempt to make contact with another species. The observations she made were stunning and are now legendary. Together with wildlife filmmaker Hugo van Lawick, Goodall documented behaviors never before dreamed of in nonhuman primates: the manufacture and use of simple tools to extract food, the hunting of monkeys and other mammals, and the ritualized sharing of meat afterward. Even the most basic observations of the humanlike long-term bond between mothers and infants, and among adult male allies, were groundbreaking stuff.

As the work at Gombe approached the end of its first decade, it was clear that this was no ordinary field study. By the late 1960s Goodall had begun to invite other young scientists to join the research effort, and each brought his or her own interest and expertise. Over the ensuing years a large assortment of Western research students and postdoctoral scientists conducted studies on every aspect of chimpanzee life. It all added up to the most intimate portrait of any species of primate ever depicted. Moreover, the students who began their work at Gombe often went on to careers in primatology themselves, training their own students and building a legacy that continues to this day. In the 1970s, the brutal killing of members of one chimpanzee community by another—so-called warfare—was first seen. In the 1980s, following a research lull brought on by the kidnapping of four students, the painstaking collection of longitudinal data began to pay off. By the 1990s, as a thirty-year record of the animals told the story of a multigenerational transmission of traditions, we began to see Gombe chimpanzees in an entirely new light. We realized they are just one of many cultural variants and began contrasting and comparing them with other long-term studies across equatorial Africa. Even Gombe itself was no longer the study of just one chimpanzee community. A second community just to the north—the Mitumba community—was observed continuously beginning in the early 1990s.

Mahale

After ceding Gombe to Goodall, Itani and his team moved just a short distance down the eastern shore of Lake Tanganyika and established a camp at the foot of the Mahale Mountains. The site would become known as Mahale Mountains National Park. In 1965, a young primatologist headed off to begin his field career, which ultimately spanned forty-five years. Toshisada Nishida is a name not nearly as well known to the West as Jane Goodall, but his contributions, and those of his many students and colleagues, stand with hers. He led chimpanzee research in the Mahale Mountains for most of his life and trained dozens of Japanese and Western scientists. Japan has a long tradition of excellence in primatology; unlike in North America and Europe, there are wild monkeys in Japan, and the scientists there cut their teeth by studying Japanese macaques. Much has been made of interesting cultural differences in the practices of primatology in the West and Japan, but the bottom line is that the contribution by chimpanzee researchers from Japan has been invaluable. Many of Nishida's otherwise groundbreaking observations had already been reported five years earlier by Goodall. When Nishida and his team saw hunting and meat eating, it confirmed that

Chimpanzee field studies of more than fifteen years. Does not include chimpanzee ecotourism sites at which some research has also been conducted.

Research Group	Location	Date(s)	Duration (Years)
Kasekela, Gombe	Tanzania	1960–	58
Mitumba, Gombe	Tanzania	1990–	28
M group, Mahale	Tanzania	1965–	53
K group, Mahale	Tanzania	1966–	52
Sonso, Budongo	Uganda	1960, 1990–	28
Kanyawara, Kibale	Uganda	1987–	31
Ngogo, Kibale	Uganda	1995–	23
Taï North	Ivory Coast	1979–	39
Taï South	Ivory Coast	1997–	21
Taï Middle	Ivory Coast	2000–	18
Fongoli	Senegal	2001–	17
Bossou	Guinea	1976–	42

* Partial data source: Langergraber, Rowney, Schubert, et al. 2014

chimpanzees are natural omnivores and that meat eating at Gombe was not an aberration. When the Japanese researchers first reported intercommunity "warfare" between M and K communities, it confirmed that the violence between communities seen at Gombe is a normal aspect of chimpanzee life rather than an anomaly.[7]

Nishida's group made one key early breakthrough for which they are not always given due credit. It took fully seventeen years, from the start of research at Gombe in 1960 until 1977, to figure out the structure of the society in which chimpanzees naturally live, and it was Nishida and his colleagues who cracked the problem. Chimpanzees were not wantonly promiscuous, as they had initially appeared; there was a community structure underlying the controlled chaos. Primatologists call chimpanzee society fission-fusion, about which we will hear much more later. Gombe and Mahale were also the first two studies of chimpanzees that demonstrated that cultural differences exist between nearby populations. Despite being only a few hours' boat ride apart, the two chimpanzee societies differ in fascinating ways that are due to their local learned traditions, not their DNA.

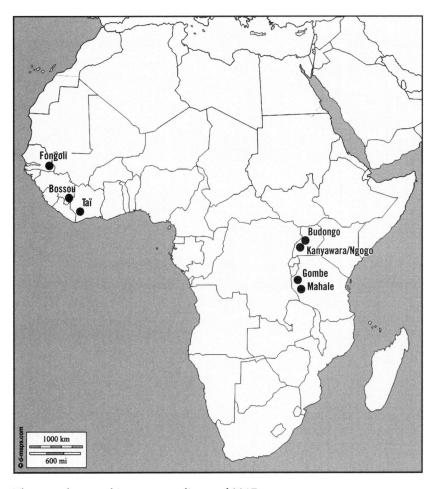

The seven longest chimpanzee studies as of 2017.

While the Gombe and Mahale studies built their foundations in the 1960s, other studies were also getting off the ground. In 1962, British primatologist Vernon Reynolds established a study site in the Budongo Forest in western Uganda, a few hundred kilometers from Gombe. His results mirrored those of Goodall and preceded those of Nishida. He stayed "only" a year (a very long study in its day), and then, as Uganda fell into a long civil war and political chaos, the site sat idle for decades, until work there began again in the 1990s under the direction of primatologist Andrew Plumptre of the Wildlife Conservation Society and later

Klaus Zuberbühler of the University of St. Andrews in Scotland. Since then Budongo has become a major source of information about chimpanzee social behavior and has become a model for integrated wildlife research, conservation biology, and the training of both Ugandan and expatriate students.

Chimpanzee Research Goes West

Until the mid-1970s, all chimpanzee research had been confined to the easternmost parts of the vast African continent. This was about to change. In 1976, Japanese primatologist Yukimaru Sugiyama established a study at Bossou, in the Republic of Guinea in far western Africa. The village of Bossou, just to the west of the foot of the Nimba Mountains, was the area in which Kortlandt had first observed chimpanzees in the early 1960s. Now Bossou again gained fame as an experimental site that used free-living chimpanzees as subjects. Unlike most research sites, in which the proximity of people and animals ruins the opportunity to study natural behavior because people are hunting the chimpanzees, the Manon people of the region treat them with respect. Run largely by Kyoto University, and in recent decades directed by Tetsuro Matsuzawa, the Bossou chimpanzees became known to science as nut crackers who use stone hammers and anvils to smash open oil palm nuts. Bossou researchers have devised a long series of clever field experiments in which they provide the chimps with tools in order to understand the cognitive elements of tool use.

As chimpanzee research spread westward, perhaps the most groundbreaking of all studies begun since 1960 was in Taï National Park in Ivory Coast. In 1979, Swiss primatologists Christophe Boesch and Hedwige Boesch-Achermann decided it was high time to gain an understanding of what chimpanzees do in the vast tracts of lowland primary rain forest that carpet the western equatorial belt of Africa. There are many reasons that the earliest phases of chimpanzee research in the 1960s and early 1970s took place in East Africa. East Africa was still British East Africa in Goodall's earliest days, which made the bureaucratic hurdles of starting work more easily surmounted. East Africa also had a safari tourist infrastructure, something that western and central Africa utterly lacked at that time. And many East Africans speak English, which made life easier for Francophobic Brits. So it took a Francophone scientist with an interest in the rain forest and in apes to head to Francophone West Africa. There he made some of the most startling discoveries in the history of primate research.

The chimpanzees of Taï National Park don't live in a landscape of broken forest and grassland, as they do at Gombe. The rain forests of the Congo basin of central Africa continue, with some gaps, through coastal western Africa, and Taï is the largest such swath remaining. Boesch's observations in the early 1980s were stunning. He saw chimpanzees collecting stones and sticks from the forest floor, carrying them to the foot of large nut trees, and using their tools to hammer open the hard-shelled treats. He observed hunting behavior just as researchers at Gombe and Mahale had, but with more cooperation among the hunters. And he showed that chimpanzees exhibit the same sort of cultural diversity across wide geographic areas that we see among people, albeit on a simpler scale.[8]

By the mid-1980s, chimpanzee research had extended to several long-term study sites, and we had begun to believe we had an in-depth picture of chimpanzee life. But in many ways, this is only when chimpanzee field research got interesting. In Uganda, Richard Wrangham of Harvard University returned to a forest that had been studied for years. Kibale National Park in western Uganda is a government-managed forest sanctuary that houses multiple chimpanzee communities. Wrangham built the Kibale Chimpanzee Project to study the chimpanzees of the Kanyawara community. It continues to this day, incorporating many researchers with varied research interests. Another community within Kibale, called Ngogo, had been studied by Michael Ghiglieri in the late 1970s and then abandoned until the mid-1990s, when John Mitani of the University of Michigan and David Watts of Yale University restarted the project. The Ngogo chimpanzee community is the largest ever recorded or studied, numbering more than 190 animals at present, triple the size of most other known communities.

These are the seven longest-term studies of chimpanzees that have been conducted as of 2016. William McGrew of the University of St. Andrews recently compiled all chimp field studies from which at least one scientific paper has been published: the list of studies short and long now numbers 120.[9] It takes many years and enormous effort to maintain a field study of chimpanzees that gets a payoff of new observations. It's a tremendous achievement by a small number of very dedicated individuals that so many studies have been carried out despite hardships of funding, local political turmoil, and the general difficulties of field research.

In addition to these studies, there are a number of other valuable but shorter field studies of chimpanzees that I will discuss throughout the book. Each has its own research angle and raison d'être. In East Africa, Kevin Hunt directed a long-term study on the chimpanzees of the Toro-Semliki Wildlife Reserve in western Uganda. These chimpanzees live on the border between forest and open country, giving us a window onto the possible habits of the earliest human ancestors, which lived, we believe, in similar habitat. Researchers including Jim Moore of the University of California, San Diego, and my former graduate student R. Adriana Hernandez-Aguilar studied the chimpanzees of the Uvinza area of Tanzania, a huge expanse of arid scrub and woodland not very far from the lusher lakeshore forests of Gombe and Mahale. Uvinza is as dry as Taï is lush, and the challenge was to follow and understand the lives of apes whose food trees were so few and far between that the chimpanzees occupied a territory of hundreds of square kilometers. This study continues today, with a new generation of primatologists doing the research.

From the mid-1990s to the mid-2000s, I studied a population of chimpanzees in the rugged mountains of Bwindi Impenetrable National Park in Uganda. The chimpanzees there have one unique attribute—they share their forest habitat with mountain gorillas. I tried to gain a better understanding of how two large apes cope with coexistence and, by implication, how perhaps our own ancestors did the same when more than one species of human roamed Africa. Not too far west, across Lake Tanganyika, Juichi Yamagiwa from Kyoto University and Augustin Basabose of the African Wildlife Foundation worked for fifteen years on the same question in a study of coexistence between eastern lowland gorillas and chimpanzees.

Farther west, in the northern part of the Congo basin, researchers Crickette Sanz and David Morgan of Washington University, Saint Louis, continue to document new behavior traditions and new aspects of ecology among chimpanzees of the Goualougo Triangle in the Republic of the Congo, where the apes coexist with lowland gorillas. Sanz and Morgan have reported new forms of tool use and obtained information on social structure and ecology. Still farther west, University of London primatologist Volker Sommer and his team have been studying chimpanzees in Gashaka Gumti National Park, Nigeria. One of the few detailed studies in far western Africa, the chimpanzees of Gashaka live in a forest-grassland mix that, like Toro-Semliki, may tell us something about how early humans would have coped with that same habitat.

Finally, exciting new discoveries have been made among chimpanzees living in the westernmost outpost of the chimpanzee range on the African continent. In southeastern Senegal, Fongoli is a mix of woodland and dry grasslands, and among its vast stretches roams a community of chimpanzees. Over the past fifteen years, a team led by primatologist Jill Pruetz of Iowa State University has recorded chimpanzees doing things that we had not suspected they would do: sitting in pools of water to cool off in hot weather (chimpanzees normally avoid even knee-deep water, and they sink like stones when placed in anything deeper) and sleeping in caves, also to avoid the baking heat of Fongoli. The Fongoli chimpanzees also invented a new use for stick tools: they ram them into holes in trees to bludgeon and disable unwitting bush babies, tiny primates that then become lunch.

I'm often asked how many studies of chimpanzees need to be done, and for how many years they need to continue, before we'll know everything important there is to know about the species. The implied question is, I suppose, what's the point of spending all the additional time and money? In reply, I usually ask the questioner how many more studies we need of human behavior before all our questions about ourselves are answered. The U.S. government spends tens of millions of dollars each year to learn more about human behavior from studies with topics ranging from adolescent sexuality to gambling habits. This does not mean we will soon have a comprehensive understanding of our own species. If I followed you around in your daily life for a few years, watching your interactions with family and friends, how well could I generalize to the human species at large? Surely with all the thousands of cultures and languages on Earth, I would at least need to spend a comparable amount of time with some other cultures, over multiple generations. That is where chimpanzee researchers find themselves today. We've compiled a small number of long-term studies, enough for some very basic cross-populational, cross-cultural comparisons. Given how long chimpanzees live (into their forties in the wild and well beyond that in captivity) and how slowly they mature (on roughly the same scale and pace as we humans do), it will take generations to sort out what chimpanzees are all about. Whether we have this time or whether chimpanzees will be extinct in the wild before then is an open question.

CHAPTER 2

Fission, Fusion, and Food

EARLY MORNING at a fig tree in Uganda, East Africa. An overnight rain followed by a misty, chilly dawn has given way to a sunny African morning. I'm sitting on a hillside that is so steep I have to jam my heels into the dirt to keep from sliding into the ravine. The mountains around me are topped with tall trees, and wherever light reaches the forest floor, the undergrowth is a thick mat of ferns. This is why the British named the place the Impenetrable Forest; these days it is called Bwindi Impenetrable National Park. Some fifty meters away at my eye level is the crown of a gigantic fig tree, the buttressed trunk of which stands in the cleft of the ravine below. I've hiked to my precarious perch to get a bird's-eye view of the action I'm expecting to unfold in the tree's crown.

Uganda has dozens of large fig species, and most provide a major feast for the animals of the forest. Many figs are dioecious—they feature both male and female trees—and local people often name the two sexes as separate tree species, further confusing an already complicated classification scheme.[1] The fruit output of this giant fig may be hundreds of thousands of centimeter-wide green figs in a season, each packed with tiny seeds. The apes love figs, and fig trees love them.

The tree stands alone, with no other fig tree within hundreds of meters. No competing trees crowd the critical sunlight. And this gigantic female tree likes it that way. She has a reproductive strategy based on forest animals eating her myriad fruits and dispersing them by excreting them distantly enough to get them out of her shade umbrella. It's a win-win-win in which she, her seeds, and the seed predators themselves all benefit.

Our fig is heavily in fruit at the moment, and it is the hub of frenzied feeding for a host of animals. In addition to chimpanzees, we've seen gorillas, L'Hoest's monkeys (beautiful black-and-white creatures with shepherd's-crook tails), parrots, hornbills, and civets, not to mention a night shift of nocturnal animals that arrive after the sun has set. These fig trees produce fruit out of cycle with one another, each taking a turn in a different month. This keeps the fruit eaters on their toes and leads to a crowded banquet when each tree's crop ripens. There are perhaps one or two per valley, and rarely is more than one in fruit at any one time. We are accustomed to trees in our temperate forests responding to spring and autumn by flushing with new leaves and a few months later dropping them more or less simultaneously. Tropical forests have rhythms too, only more subtle and less strictly seasonal. Among figs, a single tree will put out its whole fruit crop at a time when the other local figs are barren. This enables each to obtain the full attention of the seed-dispersing animals.

A chorus of loud hoots signals the approach of a party of chimpanzees, their glossy black backs all but hidden by the dense undergrowth in the ravine. They slept near the gigantic fig last night. Although they're accustomed to being watched from a distance, they won't allow us to get too close. If we do, they descend quickly to the ground and slip away. So we confine ourselves to making observations from distant vantage points using binoculars. That is fine with us; the terrain is so steep and the undergrowth so dense that standing under a tree in which the apes were sitting would provide no vantage point at all.

As the apes arrive and climb the massive trunk, they give hearty food grunts that signal their excitement that a good meal is about to begin. The big males race to the top branches to hoard the ripest fruits, which are growing in the sun. The chimpanzees are silhouettes against the sky as they sit or stand on foot-wide tree limbs, pulling handful after handful of fruit into their mouths. As the sun rises higher, we can make out individuals. Mboneire—"handsome one" in the local Ruchiga language—holds a central place in the foraging party and seems to have his pick of the fig crop. He is a jet-black male in his prime and the alpha of the community. An older male we named Kidevu—"bearded one" for his white goatee—sits next to Mboneire. Kidevu plucks figs with his one good hand; the other is withered and stiff and appears to have been caught by a poacher's snare years earlier. Martha, a big female, stands upright to reach over her head and pick figs, her baby May clinging to her back.

As the morning brightens, the chimpanzees move from bough to bough, plucking fruits. It's a good chance to watch social interactions; who displaces whom at favored feeding spots can be a signal of the dominance hierarchy. Around midmorning, the chimpanzees are looking satiated and lethargic. After two hours spent lying contentedly on the largest boughs of the fig, the party wakes up as if on cue and descends the trunk. Upon reaching the ground, they melt into the undergrowth. By their calls we can tell that the foraging party has split into at least two smaller parties, which are headed in opposite directions. The two subgroups may travel separately all day, they may merge, or they may splinter further. What began as one party of a dozen chimpanzees may finish the day in far-flung valleys in groups of twos and threes or individually. They'll likely head toward other fruiting trees, only to cycle back to this fig in a day or two to harvest the remainder of its crop. At nightfall, individuals, pairs, trios, or larger parties will climb into the forest canopy, make their beds, and go to sleep.

Fission and Fusion

By the early 1970s, Jane Goodall had made a series of dramatic discoveries about chimpanzees. They used tools, ate meat, and mothers and infants had bonds not unlike those seen in humans. But their mating system remained a mystery. Goodall surmised, based on their seemingly random comings and goings, that chimpanzee society lacked any stable unit apart from mothers and their infants. The chimpanzees would arrive every day at a forest clearing that had been turned into a feeding station for filming. Some days a given male would be alone, other days with a close ally, and still other days with a whole gaggle of community members. Females were more likely to be seen alone, accompanied only by an infant. Mating appeared to be promiscuous, apart from the struggle for dominance among adult males that gained them some degree of sexual priority. But even then, Goodall observed so many clandestine matings between young, low-ranking males and high-ranking females that the dominance hierarchy didn't seem to ensure paternity.

In those early years at Gombe, most observations were limited to the forest clearing. It was only when Goodall and her protégés began to systematically venture farther afield to follow the chimpanzees on their journeys to distant valleys that a fuller picture of their habits emerged. It became clear that they preferred a highly diverse diet; in the 1970s Richard Wrangham documented well over one hundred species of plant foods in the diet of Gombe chimpanzees (the total has since increased to well over two hundred at most chimpanzee study sites).[2] It became clear that a highly diverse, fruit-based diet was related to the chimpanzees' tendency to search far and wide for fruit trees.

At the same time, researchers were well aware that females were less sociable than males, except during the twelve or so days of their thirty-five-day cycles when males gravitated to them and vice versa.[3] During these estrous periods, a large percentage of the normally fragmented chimpanzee community coalesced around that female. Such large parties stayed together for days until the female's swelling deflated. This fluid grouping pattern was the first evidence that chimpanzees, unlike nearly every other social mammal, don't live in cohesive groups. But the inner workings of their social system remained a mystery.

A few years later, a hundred kilometers down the lakeshore, Toshisada Nishida and his team in the Mahale Mountains observed the same pattern and concluded that there was a basic structure in chimpanzee society, which primatologists call a community. The community has a well-defined range, which is hotly defended against incursions by neighboring communities. Within the community, all associations are transient except those between mothers and their infants. Nishida and other primatologists didn't realize it at the time, but chimpanzee social structure is remarkably uniform across a wide range of habits spanning the entire width of equatorial Africa. In East Africa, with its mosaic of deciduous forest and patches of grassland, we see the community or party structure. Across the vast expanse of lowland rain forest in the Congo River basin, wherever we've studied chimpanzees we find the same. And far to the west, where the rain forest gives way to more semiarid grassland and scattered forest, we see it too. The main features of chimpanzee social behavior are strikingly consistent over the whole of their African range.

Community size, however, varies widely. It can be as small as fifteen members or as large as nearly two hundred, though most known communities are in the forty-to-sixty-member range.[4] Within the community, all is in flux each day. Community members forage as lone individuals or in parties as large as fifty. Chimpanzee researchers agree that party size and its fluctuations are strongly influenced by some combination of food availability and distribution, as well as the presence of sexually available females. Other factors such as habitat quality, predator pressure, the tendency of lactating females to be alone, and the tendency of adult males to bond together for purposes of community defense may also explain why and when party sizes undergo such dramatic changes.

The Daily Grind

Finding food is a large part of the life of any wild primate. Many studies have supported the idea that primates, like other animals, live close to the borderline between energy sufficiency and energy deficit. The input of calories and nutrients must balance the output of calories expended in the search for food. A bad day foraging for food may leave an ape with an energy stress that affects other aspects of its life. Some wild primates—howler monkeys, for example—are sedentary animals that live on a largely leafy diet. Foliage might seem to be an easy food choice in the cornucopia of a tropical forest. Leaves are everywhere for the taking, and howlers eat a leafy, high-fiber diet. It's misleading, however, to think of leaves as an easy source of energy and calories. Mature leaves, darker in color and thicker in texture, are full of fiber, which is the infrastructure of the leaves' cells. They are sometimes also chock full of defense chemicals, by-products of plant metabolism that the tree has modified to protect its foliage from being eaten by herbivores. The challenge for the herbivore is selecting tender, young growth that will be easy to digest and more nutrient rich. If you're a tree, you want to rid yourself of your fruit, since that is how you reproduce. Leaves, on the other hand, manufacture your energy for you, so you want to keep them as safe from harm as possible. The combination of fiber and noxious defense compounds makes digestion a chore for an herbivore and sends chimpanzees and other primates looking elsewhere for preferred sources of energy and nutrients.

Unlike howlers and many other herbivorous primates, chimpanzees spend their days knuckle-walking their way through the forest and climbing in search of food in the canopy. Fruit provides them with quickly digested energy that they dearly need. Males may travel several kilometers per day, females typically less. The daily travel route is primarily a search for stands of trees that are in season or a particular huge fig or other fruit producer. If you want to map the daily and seasonal movements of a community of chimpanzees, the best place to start is with a map of the locations of all the fruit trees in their range. Researchers know this well. Karline Janmaat and her collaborators from the Max Planck Institute for Evolutionary Anthropology in Leipzig, Germany, studied how chimpanzees in Taï National Park, Ivory Coast, remember information about the locations of large fruit trees to guide their foraging in subsequent seasons and years. They found that Taï chimps seek out the largest trees of a given species that they can find, monitoring them for the presence of ripe fruit.[5] Their data suggested that over the course of three years, the chimpanzees approached particular large fruit trees that had been productive in previous years more often than other, smaller trees. Clearly, they remember when and where to find a good meal.

In every long-term study that has been conducted, the chimpanzee diet has been observed to be 60 to 80 percent ripe fruit. Most studies have reported two-thirds to three-quarters fruit. Chimpanzees are omnivores, including foliage, flowers, insects, and the meat of small mammals in their diet, but it would not be misleading to refer to them as ripe fruit specialists. Fruit provides chimpanzees with carbohydrates to fuel their long-distance foraging. But as we saw earlier, ripe fruit is ephemeral—it occurs seasonally and in patches that are often widely separated. The patchy distribution of fruit and its seasonal availability present major problems to hungry chimpanzees. Finding a tree large enough to feed an entire group without aggressive jockeying may mean walking for kilometers. This combined problem of quality and distribution places a premium on being able to find high-energy fruits when they're ripe while avoiding competition for them. The equation of patchy fruit sources in the diet and a fragmented, fluid grouping pattern is also seen in a few other primates, such as the closely related bonobos and the spider monkeys of the New World, both of which are also ripe fruit specialists.

Eating a fruit-filled diet impacts the chimpanzee social system in other ways. First, the diet is diverse. Few fruits are available throughout the year, so chimpanzee parties must constantly target whatever species are temporarily available. The seasonal and monthly variation in the availability of fruit influences seasonal and monthly variation in the size of foraging parties as well. This has been shown at Ngogo in Kibale National

Park, Uganda, by John Mitani, David Watts, and Martin Muller; at Budongo by Nicholas Newton-Fisher of the University of Kent and his colleagues; and by various researchers at Gombe in Tanzania, including me.[6] Fruit quality and availability may also differ from one part of the forest to another. Work at Gombe by Carson Murray of George Washington University and her colleagues suggests that individual females dominate access to preferred forest patches and enhance the status of their offspring as a result.[7] They also tend to be more aggressive when they're inside the core area of their home range than when outside, according to a further recent analysis at Gombe by Jordan Miller of George Washington University, Anne Pusey, and others.[8]

We now have detailed observational records of chimpanzee diets across Africa. In forests where the chimpanzees are not accustomed to being watched, we can use fecal analysis. We can even analyze feces or shed body hairs (recovered from nests) using stable isotopes found in body tissue to estimate diets of unseen chimpanzee populations; trees, shrubs, and grasses each have characteristic biochemical signatures. The results are similar at all sites. Watts and his colleagues compiled feeding records for the Ngogo chimpanzees and found that about one-quarter of their feeding time was devoted to figs. One species, *Ficus mucoso,* was the most important single food item in their diet. Overall, the Ngogo chimpanzees ate at least 167 plant foods representing thirty-eight plant families. Plant parts included leaves, pith, mesocarp, flowers, seeds, grasses, cambium from tree bark, and roots. As diverse as this diet is, the top fifteen plant species accounted for nearly 80 percent of the total feeding time. The chimps also ate mushrooms, honey and honeycomb, termites, caterpillars, and soil. In addition, they consumed at least ten species of vertebrate prey, including all the other nonhuman primates in the forest.[9] And in areas of human habitation, studies by researchers such as my former doctoral student Maureen McCarthy of the University of Southern California and the Max Planck Institute for Evolutionary Anthropology have shown that chimpanzees readily include invasive, nonnative species introduced by people, as well as agricultural crops, in their diet.[10]

The value of long-term data on diet is obvious; the longer a study continues, the longer the food list grows. Longer studies allow researchers to take into account years of drought and failure to fruit by key species and still understand the long-term pattern. The longer the study continues, the larger the observed home range or territory grows as the chimps explore places they rarely visit. Nevertheless, some differences do exist even between chimpanzee communities occupying nearby territories. In Kibale National Park, Kevin Potts of Yale University and colleagues showed that the diets of the Ngogo and Kanyawara chimpanzees (whose ranges are almost adjacent) overlap, with a few of the same fruit species serving as important components of the diets of both communities. But the Ngogo chimpanzees ate a less diverse diet and a more fruit-rich diet overall. They also traveled from one food tree to another more frequently than their Kanyawara counterparts. Ngogo is a moderately diverse tropical forest, but it is so productive for fruit compared to Kanyawara that the Ngogo chimpanzees may not need to diversify their diet to incorporate high-quality foods. In fact, the occurrence of *Ficus mucoso* alone, which is abundant and important at Ngogo but absent at nearby Kanyawara, appears to play a key role in Ngogo chimpanzee feeding strategies. Potts and his collaborators suggest the Ngogo-Kanyawara difference is likely due in part to the highly productive, mature forest at Ngogo, which provides an overall higher-quality diet than does the more mixed and secondary-growth forest at Kanyawara. In a separate paper, they demonstrated that the extremely fruit-rich forest at Ngogo allowed the chimpanzees to forage more efficiently as well, despite their higher population density.[11]

Everywhere chimpanzees live, they mainly eat ripe fruit. This is the case in lowland rain forest like Taï National Park in Ivory Coast and in arid grassland mixed with patches of forest at Fongoli, Senegal. Here and at each site in between, more than a hundred plant species are consumed. The chimpanzees' focus on fruit stands in sharp contrast to many other nonhuman primates that are more catholic consumers. Baboons and some other Old World monkeys, for example, eat whatever is available, including many parts of plants not eaten in the same habitat by chimpanzees. Watts and his colleagues point out that what distinguishes chimpanzees from other nonhuman primates at Ngogo is that chimps continue to search for preferred fruits even when they're scarce, whereas the local monkeys—red-tailed and blue guenons and gray-cheeked mangabeys—cope with fruit scarcity by switching to other foods. Chimpanzees simply invest the energy to forage farther and wider to find their cherished figs and other fruits.

Diet and feeding have been particularly well studied and analyzed at Ngogo and Kanyawara, but similar evidence exists at Gombe, Mahale, and Budongo in eastern Africa, and Goualougo in central Africa to support the notion that chimpanzees are, at heart, ripe fruit specialists. Frugivores, primate and otherwise, tend to share a suite of behavior patterns. They travel far and wide in search of scattered fruit trees, and once they find those trees they have to compete for fruit with others in the group. The reward is a high-quality, carbohydrate-rich, high-energy food. Travel patterns are closely linked to dietary quality and, therefore, fruit trees. In Taï National Park, Antoine N'guessan and colleagues found that chimpanzee parties cut back their daily travel distance during times of fruit scarcity, while Simone Ban and coauthors reported that the same chimps oriented their travel in the direction of particular trees that held high-fat fruits.[12] There is limited evidence for chimpanzees switching to fallback foods during periods of scarcity. Wrangham and colleagues showed that chimpanzees and other primates in Kanyawara eat low-quality but readily available foods such as leaves and bark when necessary. They may also change party size, as shown by Wrangham and also by Rebecca Chancellor of the Great Ape Trust and colleagues, or they may travel to parts of their range in search of new fruit trees.[13]

Newton-Fisher and colleagues examined the roles of fruit abundance and distribution among the Sonso chimpanzee community at Budongo. If competition for fruit has led to the fragmented, fluid party-foraging system, do periods of fruit scarcity and abundance correspond to fluctuating party sizes? In fact, they do. While party sizes did not vary according to Newton-Fisher's estimates of food abundance across the entire community range, they did correspond to the varying sizes of patches of food—mainly the occurrence or lack of stands of trees with ripe fruit. It wasn't all about sheer abundance of food; the Sonso chimps expanded the number of food patches in which they foraged as food became more abundant, and they did not always form larger parties to do so.[14]

Chie Hashimoto of the Primate Research Institute of Kyoto University and colleagues compared the relationship between fruit and party size at four sites across eastern and central Africa. They showed that party size corresponded strongly with fruit abundance. But paradoxically, seasonal changes in party size did not correspond to seasonal changes in fruit abundance. A factor other than fruit influenced party size changes from one season to the next. That appeared to be a social factor, perhaps one that is decoupled from the availability or abundance of fruit. Hashimoto and colleagues theorize that fruit abundance influences party size seasonally only in forests in which fruit is so limited most of the year that party sizes tend to be small overall. Then, in months in which fruit becomes abundant, it is not the fruit but rather social factors such as swollen females that generate larger parties. The impact of fruit is therefore real but easily overestimated. The Hashimoto study was notable in disagreeing with others about the role of fruit in party size. It was also notable in directly comparing the nexus of fruit and party size across multiple study sites.[15]

So the quality, availability, and distribution of plant foods, in particular fruits, strongly influence how and where chimpanzees eat, how far they travel each day, and how social they are while searching for food. But there is clearly another powerful influence: a social factor. We now need to examine the link between diet and social behavior, something that has puzzled and excited primatologists for more than fifty years.

Food, Females, and Social Structure

A half century ago, a few enterprising ecologists attempted to make sense of the diversity of primate mating systems by linking them to the local environment. Some of these were the leading lights on animal ecology and evolution of their day: John Hurrell Crook, Steven Gartlan, Paul Harvey, and Timothy Clutton-Brock. A series of papers compared primate mating systems to their diet and made the case that primate societies evolved in direct response to what they ate. They cited an apparent tendency for species with polygynous mating systems (in which one or more males mate with many females) to eat a fruit diet. This, they argued, meant that polygyny and fruit were linked.[16]

These sorts of one-to-one correlations turned out to be highly simplistic. As more and more primate species were studied, it became clear that their mating systems did not fit neatly into any dietary category. By the 1980s, primatologists had turned to empirical studies of the competing influences of food and sex on group living in an attempt to solve the puzzle of primate sociality. Some studies indicated that primates living in larger groups suffered reproductively, presumably because of the additional mouths competing for food, versus those in small groups. Wrangham argued the opposite—that in some circumstances, the goal of occupying and defending a fruit tree might place a premium on groups with more members. In this case, the larger the group, the more food each individual could obtain. Wrangham was trying to explain the unusual chimpanzee fission-fusion social system, in which males spend their entire lives in their natal community, while females migrate. It turned out that this pattern is less unusual than it appeared to be in the early 1980s when Wrangham framed his theory. Nevertheless, it provided a strong platform for testing theories in the 1980s and 1990s about the inner workings of primate societies.[17]

In many primate species, females are the core of the social group; the members of this nucleus of mothers, daughters, sisters, and grandmothers remain together their entire lives. They defend their food sources and the territory itself from other groups. Other species in which males remain in their natal group for life feature powerful male coteries that control access to food patches such as trees full of ripe fruit. These males have an incentive to cooperate, or at least to refrain from too much competition. They can achieve great things by working together that would be impossible individually. Male coalitions in chimpanzees also carry out Shakespearean conspiracies to obtain females and power.

So species in which females rather than males migrate feature a cadre of males who form the infrastructure of the group. These often include brothers and cousins. A chimpanzee community is structured exactly this way, with females migrating between communities while males occupy and control territory. Because male chimpanzees remain in the communities in which they were born while females migrate, we expected that most of the males would be related. This kinship would provide them with a genetic incentive to cooperate in border patrols, hunting, and the like. But several studies have shown that the degree of relatedness among males is not especially high. We now think that male cooperation is based mainly on the shared benefits of working together, with kin selection playing some role as well. We still don't understand exactly why this system evolved in chimpanzees, but there is suggestive evidence that it may be the deep pattern of all the great apes, and perhaps of ancient humans as well. The evolutionary history of any primate plays a profound but poorly understood role in the form the mating system takes in the present day. This history is difficult to tease apart from the ecological influences on the social system.

The modern Darwinian approach to understanding primate social behavior and mating systems offered a stark upgrade over the broad-stroke ecological comparisons of the past. By considering the selfish interests of each individual and how these differ between males and females, we can reconstruct some elements of how the chimpanzee grouping pattern came to be. The mating strategies of male and female chimpanzees are at odds, even though their goal of reproductive success is the same. Each sex tries to maximize the odds of producing offspring that will thrive and live to sexual maturity to become breeding adults themselves. For females, that means optimizing access to good food in order to carry a healthy fetus to term and then rear and nurse the infant successfully for years before weaning it and getting back to the business of reproduction. For males, it means mating with as many females that are likely to be ovulating as possible while preventing rival males from mating with those females.

Food is a more consequential link between behavior and reproduction in females than it is in males. This is why scientists refer to females as the ecological sex and most theories about the evolution of mating systems in social mammals are built around females. Males are typically less dependent on food sources. Instead, their social groupings tend to be patterned to maximize their access to females. Male chimpanzees form coalitions to defend territorial borders and attack enemies. They defend territories on which their females live and also try to control females directly, even coercively. But females still hold some sway. As with other mammals, females don't need to worry about finding males with whom to mate; males will find them. Females spend their days in search of ripe fruit trees, and they're not especially eager to share the bounty of those trees once they've found them. Since males must map themselves onto females, in months of fruit scarcity they may be trying to maintain access to females over a wide area.

Males band together to attempt to control those females, and they hotly contest territorial boundaries, over which a neighboring community's males might intrude. This, more than the joint defense of fruit trees, is likely to account for the form of the chimpanzee social system. Females prefer male company only when it suits them—while they're ovulating or when males have located food trees or captured prey that females want a share of. The general trend among primates is that the sex that disperses tends to be the less social sex, and this is certainly true for chimpanzees. Studies of East African chimpanzees in Gombe, Mahale, Budongo and Kibale have shown females to be far less social than males. It might be that in East Africa, female chimpanzees are more attached to their territories or the core areas within them than they are to males or to the community at large. This would account for the observation at Gombe and Mahale that some females who are long-term members of the community remain peripheral, only joining parties when those parties venture into their own foraging ranges.[18] Female core areas tend to be far smaller than the territory monitored by the males, which encompasses the entire community range.

Historically, most chimpanzee research has focused on males. They're more sociable and they tend to be bolder early in the habituation process. There may also have been an entrenched bias among scientists that "what males do is more interesting." Female gregariousness is to some extent a function of the size of the community. The more chimpanzees there are, the more likely a female will spend time with another chimpanzee. But Julia Lehmann and Christophe Boesch of the Max Planck Institute for Evolutionary Anthropology showed in a series of papers that among Taï chimpanzees, females were only slightly less sociable than males. Female bonds with one another were as strong as those between males and females. Whether this striking difference between Taï and other sites is due to unusual circumstances at Taï that encourage cooperation among females or is a geographic difference between eastern and western chimpanzees in general is unknown. As the size of the Taï community declined drastically through disease and poaching, sociability actually increased. Party size and the duration of time that chimpanzees spent in parties increased, leading Lehmann and Boesch to conclude that smaller communities are more cohesive and that males and females also have stronger bonds in smaller communities.[19] Indeed, in the largest community known, Ngogo, chimpanzees are more likely to be seen alone than in other sites, perhaps simply due to the sheer number of community members foraging at any given time.

Gombe may in some ways be the anomaly. Gombe females spend a third more time alone than females in any other study site. Nearly half of all Gombe females do not migrate at sexual maturity, contrary to the pattern observed in other chimpanzee studies. Livia Wittiger and Boesch found that Taï female gregariousness may be determined by both the distribution of fruit and the number of females with swellings at a given time. When fruits were available in dense clusters, such as when stands of trees ripened simultaneously, females became more sociable. Taï females were far more likely to be in parties of all females than were females in other chimpanzee communities.[20]

Putting Food and Female Behavior Together

While the abundance and availability of food is usually but not always associated with larger party sizes, the presence of estrous females *always* results in larger parties. Males want to be near females that may be ovulating. This includes males who, based on paternity data and observations, have nearly no chance of fathering a baby with popular estrous females. During my research at Gombe and also in the longer-term data analysis there, party size actually increased in the dry season, the lean time of year when most researchers had documented weight loss among the chimpanzees. It turned out, however, that during my own study this may have been a function of my focus on hunting parties and didn't necessarily reflect parties as a whole. Monica Wakefield of Northern Kentucky University analyzed female sociability at Ngogo and found that females did form associations that went beyond those we've typically associated with female chimpanzees. In addition to joining parties, they formed cliques that actively attempted to stay in contact with one another.[21] Such female gregariousness may have been abetted by the extraordinarily productive forest at Ngogo, which mitigates the intensity of competition among females. This belies the commonly held belief about the lack of sociality among female chimpanzees, and it gives Taï and Ngogo something else in common.

In the Mahale Mountains, Akiko Matsumoto-Oda and her colleagues at Kyoto University found that parties of one to five chimpanzees were most common; they accounted for nearly 95 percent of all male parties. Parties were on average largest during the two rainy seasons and smallest during the two dry seasons. These periods of largest party size coincided with maximum fruit eating. The researchers also found that party sizes were larger when sexually swollen females were present. A strong correlation among party size, fruit consumption, and swollen females is common to most field studies.[22] Untangling which of these effects is stronger is the challenge.

Food availability has a stronger effect on party size in some communities than in others. Wittiger and Boesch argue that this is because important fruit tree species vary in density and abundance, which can have a powerful effect on the association patterns of the chimpanzees, and therefore their gregariousness.[23] When fruit is scarce—for instance, at Gombe for several months during the long dry season—the chimps disperse widely, their body weight declines, and they forage in small parties or alone. They give food calls less often too—perhaps to avoid drawing unwanted attention when they find a tree with ripe fruit. It's clearly a difficult time, and Gombe may experience a more severe and prolonged dry season than Taï and other sites. The sum of this indirect evidence that fruit scarcity affects Gombe chimpanzee social grouping may reveal key differences between Taï and the other chimpanzee studies about the effect of food on party size and grouping patterns.

Nests

Every night, a chimpanzee builds a bed and goes to sleep. These nests are nearly always made in the crown of a tree. Occasionally a chimpanzee will nest near the ground and, even less often, on the ground. Mothers with infants take their babies to bed with them; otherwise, each member of the community sleeps alone. The significance of nests for understanding chimpanzee society is that they are a highly visible signature of the use of space. A researcher can estimate the number of chimpanzees in a given forest by counting nests. By using some statistical corrections, one can gauge the age of nests and thereby estimate the number of new nests being made each week. Presumably, the earliest apelike humans made and used shelters at night, so understanding the ecological and cognitive aspects of nest building among chimpanzees may provide some insights into the behavior of early hominins. It's interesting that some apes build fairly elaborate nests, while others—such as orangutans—build none.

Gorillas, for example, typically build enormous, simple bowl nests on the ground wherever the group stops for the night, although in my research site in Bwindi Impenetrable National Park, Uganda, they sometimes nested in trees.[24] Sometimes great apes also build nests during the daytime when they nap, but these are lightly constructed, flimsy affairs compared to their night beds.

Primatologists mostly ignored nests and nest making for decades. Even in Goodall's 1986 opus on Gombe chimpanzees, nest making is mentioned only in passing. But recently, chimpanzee researchers have turned to understanding the finer details of their construction and use. We now know, for instance, that Taï chimpanzees remember the locations of many fruit trees in their habitat and tend to make their nighttime nests in proximity to them, presumably to provide for a convenient breakfast the next morning. Chimpanzees in Toro-Semliki Wildlife Reserve and at Bwindi select particular tree species as nesting platforms. At Semliki the choice is based on tree structures that are optimal for supporting a sturdy but comfortable nest. R. Adriana Hernandez-Aguilar of the University of Oslo showed that chimpanzees living in arid areas with low tree density make use of the same trees and nests multiple times, something chimpanzees elsewhere very rarely do.[25]

Although tree nests are the norm, in some forests chimpanzees make nests on the ground. In Bwindi Impenetrable National Park, chimpanzees make ground nests on steep hillsides. They're constructed with the lower end built on top of a fallen log or other forest floor debris. They have the appearance of a poorly made but level human campsite. In other parts of Africa, ground nests are more common. Kathelijne Koops of Harvard University and her colleagues studied nest building among the chimpanzees of the Nimba Mountains in Guinea, West Africa. They found, as we had, that chimpanzees indeed choose particular trees, or at least those with particular architecture, for their nests. They choose tall trees at elevations above one thousand meters with low-growing, large branches. Koops and her colleagues tested various hypotheses to explain nesting patterns and concluded that they served a thermoregulatory function that also avoided maximum moisture and humidity; nests were higher in trees and at higher elevations during the warm, wet season. She and her collaborators also showed that males were more likely to build nests on the ground, perhaps related to their more massive size.[26] At Gombe, an alpha male named Freud sometimes slept overnight on the ground, occasionally without even making a nest, usually at the bottom of a tree that held a particularly desired swollen female.

David Samson of the University of Nevada, Las Vegas, and his coauthor Kevin Hunt showed that chimpanzees at Semliki select their nest trees carefully: a species that composed less than 10 percent of the trees was used to make nearly 74 percent of the nests. The chimps built their night nests strongly, choosing limb diameters and tree architecture that provided solid biomechanical support. The authors felt that the preferred tree species was clearly chosen for nesting because its boughs offered firm and stable platforms for nest building.[27]

When chimpanzees stop traveling after a long day, the party climbs into selected trees, each makes a nest, and they bed down individually. At that point the researcher puts away her notes and heads back to camp. The chimpanzees spend the next twelve hours unobserved, and interesting things may happen during the night. Occasionally we note that a chimpanzee wakes up in the morning in a nest that is not the one he went to sleep in. Either he ventured out for some reason and then made a new nest in the dark or he displaced someone else from a nest and took ownership. These instances are rare, but nocturnal observation is so infrequent that we can't really estimate how often it happens. There are anecdotal observations from Goodall and others of chimpanzees spotted while traveling across farm fields in the light of a full moon. Koichiro Zamma of the Great Ape Research Institute in Okayama, Japan, studied sleeping and waking cycles among wild chimpanzees in Mahale Mountains National Park. This is practically the only study ever conducted of the night life of wild great apes. He found that the chimps woke up often during the night, usually due to the same things that wake us up: the sound of a nearby animal or possible intruder or the need to urinate or defecate.[28] In my study of chimpanzees and mountain gorillas in Bwindi Impenetrable National Park, our camp was located at the mouth of the valley in which our animals frequently nested. Sounds wafted down the valley with the breeze, and we heard chimpanzees or gorillas calling nearly every night, sometimes for hours.

Sleeping in a nest is not essential to a chimpanzee's well-being, and yet they do it habitually. In captivity they are provided with nesting materials, and if they are not, we feel they're being deprived of a basic necessity.

Nests no doubt provide comfort, but they also provide some predator protection and shelter from the elements. Fiona Stewart of the University of Cambridge and a few other researchers believe nests serve an important function of protecting chimpanzees from parasites, which might include malarial mosquitoes. Stewart actually slept out in a chimpanzee habitat at night once a month, both on bare ground and in nests that she or her chimpanzees had constructed. She found her sleep was more disrupted and less deep when on bare ground.[29] So sleep quality may be the bottom line for wild chimpanzees when building a nest.

Calling Cards

Vocalizations play a key role in chimpanzee social life. Although chimpanzees display a variety of calls given at close quarters, long-distance calling is a prominent feature of chimpanzee society and its connection to food. Because of the fragmented nature of the chimpanzee fission-fusion society, individuals are often out of visual contact with one another. Pant hoots are a way for them to stay in contact as they forage separately or in small parties. Researchers since Goodall have argued that pant hoots may function to identify a caller to other, more distant chimpanzees. According to a recent study by Pawel Fedurek of the University of Neuchatel and colleagues, pant hoots may identify the caller's social status, age, and individual identity.[30]

Wailing pant hoots are given in a variety of contexts, sometimes accompanied by loud drumming with the hands or feet on tree root buttresses. They appear to be crucial in organizing the movement and meetings of foraging parties. When traveling chimps pant hoot, they often set off pant hoots or choruses of them from distant parties. At times a quiet valley that appeared to be empty of chimpanzees suddenly erupts in pant hoots, followed by more distant ones. Fedurek and colleagues studied the role that pant hoots play in the social behavior of males at Kanyawara in Kibale National Park, Uganda. They found that these long calls were good indicators of strong bonds among the callers. As with grooming, you can tell who is allied to whom, and which bonds are strongest, by measuring the pattern of pant hoot chorusing. Fedurek and colleagues concluded that pant hooting plays a complex but key role in regulating social interactions within and among traveling parties.[31]

We know that pant hoot calls vary from one population to the next. Is this comparable to human dialectical differences? Adam Clark Arcadi of Cornell University and his collaborators believe so. But Mitani and his colleagues argue that body size and aspects of the forest environment mold the acoustical structure of the calls of geographically distant populations.[32] I was once following Frodo, the enormous Gombe male, as he traveled alone up Kakombe Valley. A chorus of pant hoots erupted from above the waterfall ahead of us. I had many times seen Frodo answer those calls rather casually, or ignore them, but in this instance he jerked his head in the direction of the calls, then raced off in their direction, scaling the cliff face next to the waterfall. I took the long route around the falls, and when I caught up to him fifteen minutes later, he was eagerly hunting red colobus monkeys with the other males. It seemed clear that something in the content or the context of that pant hoot chorus had conveyed a message to Frodo that something was happening that he'd want to be a part of. Acoustical analysis of the calls he heard might have shed light on whether something in their content provided meaning for Frodo. More likely, however, is that the context—decades of traveling those trails and hearing the same voices calling to him in myriad circumstances—told him what he needed to know.

Pant hoots are given by chimpanzees when two parties approach one another in the forest. They are also given during displays, either when alone or with a chorus of others. And they also may represent announcements or inquiries, such as when a male chimpanzee arrives at a particular place and drums on the flaring buttresses of the surface roots of a large tree. Clark Arcadi and his colleagues studied drumming among Taï males and found that low-frequency drums, audible by human ears up to a kilometer away, differ in tempo and cadence among males and may serve to identify individuals to each other. They also found differences between Taï and Kanyawara in drumming styles and patterns, which are presumably cultural variants between the populations.[33]

If the meaning of pant hoots is not entirely understood, other calls are more obviously related to daily concerns in chimpanzee life. Food calls are often given when chimpanzee parties arrive in a tree holding ripe fruit. These grunts can sound contented and relaxed or, when the chimps are particularly hungry or the food is especially relished, frenzied. It's hard not to imagine the apes are expressing their delight at discovering a wonderful meal waiting for them. When the fruit bounty is large, chimpanzees may be more vociferous about their discovery. In captivity, chimpanzees have calls that indicate not only that food is present but also whether the bounty is large. Amie Kalan and her collaborators at the Max Planck Institute for Evolutionary Anthropology found that Taï chimpanzees altered their food calls with regard to the size of the fruit tree and also with respect to the particular fruit species they'd located. They gave higher-pitched calls when they discovered highly valued fruits. The researchers could not ascertain whether these calls had an effect on other chimpanzees who heard them, but they felt sure that the call differences reflected more than just emotional arousal. They were *intended* to communicate something important.[34] Meanwhile, Budongo male chimpanzees give food calls mainly when a close social partner or ally is within earshot, according to Kate Slocombe of the University of York and her coauthors. They also modulate their call production based on how much food is available and whether it can easily be monopolized. In other words, they appear to judge whether they've found more food than they can monopolize, and, if so, they are more willing to share with other members of the community.[35]

Alarm calls are also part of the chimpanzee vocal repertoire. Like other calls, they indicate an ability to understand and influence the mental state of other chimps around them. But alarm calls, because they're given when danger looms, are more central and critical to chimpanzee well-being. The "wraa" is an alarm call given in response to dangers such as leopards or pythons, or sometimes surprise appearances by strange chimpanzees. In field experiments, Catherine Crockford of the University of St. Andrews and colleagues presented Budongo chimpanzees with a threat (an artificial snake) and examined the alarm call response. They found that alarm calls were given mainly when the caller was traveling with members of the party who had not seen the snake or who had not been present when the first alarm call was given. In other words, Budongo chimpanzees use alarm calls to inform naïve group-mates of danger.[36]

There is a great deal of content in calls that we are just beginning to understand. When a chimp is attacked by a member of his or her community, the vocal response is often a scream, a "waa" bark, or both. In another of Fedurek's innovative field studies of calls, he and colleagues studied both calls and found that the victim of an attack was likely to counterattack after giving a waa, but not after screaming. This suggests that the function of waa-barking is to warn the attacker that the victim is going to fight back. Screaming, meanwhile, appeared to serve to recruit or try to recruit support from others in the vicinity rather than signal an immediate counterattack by the recipient.[37]

Why Fission-Fusion?

Chimpanzee social structure is unusual. It's a product of some combination of food availability and distribution and access to females, with a dose of evolutionary history as a template. Quantifying the two variables we can most easily examine—food and females—leads us to the view that female chimpanzees disperse themselves on the landscape of an African forest in order to maximize their access to ripe fruit and minimize the number of competing mouths around them. This equation only changes when a female has a sexual swelling and become more sociable, as males also gravitate to her. The dual factors of food and female reproductive cycles dictate party sizes, which fluctuate very much in tune to those influences. But as important an influence as food is, it appears that the availability of females trumps the availability and distribution of food. Other, unappreciated factors may be at work, but they are either difficult to measure or don't occur in all chimp communities. Some researchers have questioned whether we fully understand the structure of chimpanzee communities. This doubt is based on the sometimes uncertain position of females in a community. Some females seem to be peripheral, both literally and figuratively, appearing even after many years of habituation only at certain times and places.

Questions remain about male social behavior too. Males are more gregarious than females. However, Kevin Langergraber of Arizona State University and colleagues working at Ngogo did not consider males to be any more sociable with other males than one would predict by chance association. So the old idea that a chimpanzee community has at its heart a cohort of males may be true but without some of the implications assumed by past researchers. The same applies to female alliances; when one controls for the overall lower sociability of females at Ngogo, they associate more frequently than one would predict by chance.[38] So the long-assumed pattern of male alliances and independent, less gregarious females may need revising.

The most poorly understood link between social structure and individual behavior is cognition. We have made dramatic strides in studies of the minds of great apes and other animals over the past twenty years; nearly all the studies have been done in captivity with laboratory or zoo chimpanzees. Studying the mind is far more difficult under natural conditions, where circumstances can't be manipulated to experimentally study the thought process through the reaction of chimps to various stimuli. It's essential that in the coming decades the next generations of primatologists do exactly these sorts of studies. Chimpanzees make use of much the same senses of smell, hearing, and sight that we humans possess. This removes the barrier that makes understanding the world from the perspective of a dog, with its hypersensitive olfaction, or a dolphin, with its echolocation, so much more difficult.

It took seventeen years to advance from Goodall's original impressions that chimpanzees lived in amorphous societies to Nishida's realization that distinct social units exist. Since the model of community structure with fluid foraging parties was established, it has taken decades more to parse out the relative environmental and social influences on it. We now have enough comparative information from a variety of study sites to say with certainty that local differences in social structure are fairly small and are influenced by some combination of differences in demography and the forest itself. In comparison to many other well-studied primate species, chimpanzees are remarkably uniform in their social behavior across a wide range of habitats. Whether we observe them in lowland rain forests or in arid grassland–forest mosaics, they exhibit essentially the same community structure with the same party system. Given the complexity of chimpanzee behavior, and the complex array of influences on it, the uniformity is remarkable. What remains is to better understand how chimpanzees see the world, and how their intelligence serves them in coping with the social and ecological complexities of their lives.

CHAPTER 3

Politics Is War without Bloodshed

Luit circles Yeroen with all his hair on end at a distance of about 15 meters. He stamps his feet and thumps the ground with the palm of his hand. He picks up any sticks or stones he finds in his path and hurls them away. Yeroen is sitting on the grass and glances swiftly up at Luit every now and again. When the challenger is behind him Yeroen does not turn around, but moves his head slightly so that he can unobtrusively watch what Luit is doing through the hair on his shoulders. Sometimes, Luit comes within a few meters of Yeroen. Yeroen then stands up, hair on end, and takes a step forward. During this brief confrontation the two males do not look at each other, and, as soon as Luit has passed by, Yeroen immediately returns to his spot on the grass.

Frans de Waal, *Chimpanzee Politics* (1982)

THIS CONFRONTATION, in which alpha male Yeroen is challenged by a rival in a zoo in the Netherlands, depicts a battle in the early stages of a long, dramatic contest over alpha status. But it's far more than a one-on-one battle. Dominance is, in our popular imagination, a quality associated with leadership that is part and parcel of an individual's personality. But in nonhuman primates, we believe that dominance is an emergent property in a relationship, rather than a thing unto itself. Dominance relationships arise between animals based on repeated interactions over long time periods. Historically, dominant members of a primate group were understood to be animals that were, for example, groomed by others more often than they themselves groomed, animals that attacked others more often than they themselves were the victims of attacks, or animals that received the support of others in fights while rarely offering their support in return. Dominance is a power imbalance. Indeed, a primate's dominance rank is often assigned by researchers on the basis of grooming received versus grooming given. So we can cautiously assign dominance rank to animals after long-term observation, being aware that other factors such as the directionality of pant grunts, addressed later, also play a critical role in understanding status.

Social interactions are typically used to sort animals into a linear hierarchy—a pecking order, so named because of an early study of dominance in chickens. But there's a problem: dominance relationships are rarely strictly linear, and they're not necessarily limited to one-on-one interactions. The fallacy of the pecking order can be seen in the earliest field studies of primate dominance. Researchers placed peanuts between pairs of male baboons, hoping to learn the dominance dynamic between them. One animal or the other always took the peanut, and this was duly noted. But there was a self-fulfilling prophecy at work. By providing a forum for males to test their relative ranks, which were perhaps undefined at the time, the researchers may have created, or at least reinforced, the pair's dominance relationship.[1]

So what exactly is dominance? Is it as simple as being able to take a peanut away from another member of the group? Or perhaps it's being able to mate with the female of your choice? Priority of access to mates is a key outcome of dominance and is on display in chimpanzee society at times, but it is certainly not the entire story. Dominance is about asymmetry in chimpanzee power relationships. Rarely is the hierarchy completely linear. The alpha may dominate the beta, who in turn dominates the gamma. But the web is typically more complex than that, because coalitions, alliances, friendships, and a great deal of Machiavellian manipulation of others come into play.

Dominance in chimpanzees is not about physical size. Each sex has a dominance hierarchy, and all adult males are usually dominant to all females. Adolescent males do not typically exhibit the hallmarks of a dominance struggle, according to research by Aaron Sandel and colleagues at the University of Michigan, Ann Arbor.[2] They don't submissively pant grunt to one another, and dominance among them is hard to discern. But once a young male reaches adulthood, he begins his climb to higher rank by taking on and intimidating each adult female. When he has risen in status above the most dominant female, the young male finds himself at the bottom of the male dominance network. Only time and repeated jousts with higher-ranking males will determine his ultimate highest status. Hogan Sherrow of Southern Oregon University found that dominance relationships among males at Ngogo were already well established by the adolescent years.[3] Some males are inveterate social climbers, cleverly serving their own ends by ingratiating themselves with high-ranking males and females. Others rely more on brute intimidation, which does not necessarily carry the day. And then there are males who seem to care little about their social status and are content to live out their lives on the edges of the struggle.

Dominance manifests itself in many ways in chimpanzee society. It's visible in body language and in the use of personal space in ways small and large. We see it in tactile communication, vocal communication, male competition over females, and female choice of males. Dramatic dominance displays, in which males race, with hair bristling, through the forest, snapping sapling trees and tossing logs as they go, make good television documentary fare but are actually not very common. Even the kind of coalitionary dominance behavior that de Waal wrote about happens mainly during periods of social upheaval.

What It Means to Be Alpha

"Alpha male" has become a part of vernacular English. The phrase has both good and bad connotations. If Steve Jobs is described as having been the alpha male of the personal computing revolution, the term suggests leadership and authority. But when you describe some guy at a party as an alpha male, it can mean he's an overbearing jerk who uses his dominant personality in an abusive way. That love-hate relationship that we have with alpha males extends to chimpanzee society. Once a male reaches alpha rank, he changes. He spends his entire day in leadership mode, his hair often standing on end to make him look more impressive and impregnable to challenges. Behaviors that may have helped him rise in rank, such as currying favor with the right allies, are dropped in favor of behaviors that help him maintain his hold on power. The most famous of all alphas in recorded chimpanzee history, the long-reigning Mahale alpha Ntologi, shared meat liberally as he rose in rank. But Toshisada

Nishida showed that once he had achieved alpha status, his generosity dropped, and he began sharing meat mainly with those whose political support he still needed most.[4]

Alpha males tend to reach that lofty status by the time they are young adults between their late teens and early twenties, or else not at all. There are notable exceptions. Goblin at Gombe became alpha male at fifteen, barely sexually mature, and was alpha (with intermittent breaks due to successful challenges from others) through his midtwenties. Kamemanfu at Mahale became alpha at the ripe age of thirty-nine, and he held the rank until age forty-three. The tenure of an alpha can last for a few months or many years; Ntologi's sixteen-year reign is the record.[5] Knowing the average length of a male's tenure won't tell us much because of the extent of variation between and within chimpanzee communities, but the most commonly recorded alpha tenure is about four years.

Good Grooming

The most obvious way dominance is asserted by male chimpanzees in daily life is through their body language. A high-ranking male approaches a low-ranking one and, as he reaches the subordinate's proximity, the lower-ranking chimp gets up and moves away, allowing the dominant animal to pass by or to sit in the vacated spot. Because the subordinate usually moves before the dominant has reached him, to a casual observer the two acts might seem unconnected. The social life of chimpanzees is about such subtle displays of high and low social status.

Grooming is both an important currency of dominance and a signal of status for observers. Grooming reciprocity—or lack of it—is the basis for primatologists to assign rank to social primates. An alpha is groomed often by many group members; a low-ranking chimpanzee does far more grooming than he or she receives. Chimpanzees spend a fraction of each day grooming, and the pattern follows lines of kinship and dominance. A recent analysis at Gombe by Joseph Feldblum of Duke University and his collaborators found that the number of grooming partners was the key variable in assessing rank. Males rose in rank by establishing grooming relationships with many other males rather than by investing in intense grooming relationships with any particular individuals.[6]

Unidirectional or imbalanced grooming is more easily understood than mutual grooming. Low-ranking animals usually "groom up," seeking to solidify much-needed alliances with higher-ranking group-mates with an effort that is very low cost on their part. They don't receive full reciprocity for their actions, but even limited reciprocity may benefit them; beggars can't be choosers. Mutual grooming, on the other hand, may strengthen long-term bonds as well, although a recent study by Zarin Machanda of Harvard University and her coauthors suggested that it mainly serves to prolong a grooming bout without necessarily securing longer-term benefits.[7]

Grooming is part of what John Mitani referred to as the "enduring and equitable" social network among male chimpanzees. While many social animals form temporary bonds to serve their immediate needs—finding a short-term partner for help in a fight, for instance—not many species form enduring bonds that last a lifetime. Male chimpanzees, because they remain in the community in which they grew up, have an obvious benefit to gain from long-term bonds. They also have the intelligence to remember relationships for decades, whether they are with mating partners, close allies, or bitter rivals. Mitani found that the stability of male bonds relied on kinship and the quality of social relationships, the latter defined by the web of grooming partners. Unsurprisingly, grooming was very much based on dominance relationships, as measured independently, but kinship also played a key role. Maternal half siblings—same mother, different fathers—groomed more equally with one another than did males with their nonrelatives, and they also maintained longer-term social bonds. As one would expect in a species in which males remain in the same area their entire lives, brothers had the longest-lasting and most equitable grooming relationships.[8]

Male bonds are reinforced daily through mutual grooming. Males tend to groom and be groomed more frequently than females, and to have a larger number of grooming partners than females do. This may be because females rarely groom one another, while males groom both females and other males. Grooming can be such an important factor in climbing the dominance hierarchy that males might conceivably compete for opportunities to groom and be groomed. Kate Arnold and Andrew Whiten of the University of St. Andrews found that in Budongo Forest, male chimpanzees groomed up the dominance hierarchy, and males adjacent to one another in social rank groomed together more than males of very different rank. Being at or near the top of the dominance hierarchy enabled Budongo males to enhance their mating opportunities, even with females whom they didn't groom.[9]

David Watts has a slightly different take on grooming strategies in relation to dominance rank, based on his research on the Ngogo chimpanzees. At Ngogo, the large number of males in the community creates myriad grooming opportunities and a complex matrix of male social interactions. As one would expect, the higher a male's rank, the more grooming partners he has. The grooming reciprocity flows from low-ranking to high-ranking males, with males low in the hierarchy grooming far more often than they are groomed. But grooming is concentrated mainly among males close in rank, as found in Budongo and elsewhere. Watts concluded that the number of available males on hand to groom and be groomed likely has a major impact on the strength of the effects he reported.[10]

Not all studies have found a relationship between grooming and dominance, or between grooming and the closeness of rank order among males. In the M community in Mahale Mountains National Park, grooming depended on the numbers of partners in the community, but without much of an association between grooming and rank.[11] Similarly, an early study at Gombe by David Bygott found that the age of the males was more important than their rank. Seniority as defined by age was a determinant of which males were groomed most often.[12]

Shakespearean Male Tales

In the early history of primate field studies, a disproportionate amount of attention was paid to males. Baboons, the earliest subjects of primate field research, are big bodied and live in large social groups. They're easily accustomed to human observers and they're abundant in many of the game parks in East Africa. Males are full of bravado and display every bit of it to other males right in front of enthralled observers. Mothers and their infants are far less confiding, retreating to the security of thickets. When the primatologists of the 1950s and early 1960s watched baboons, they focused more on males than females for exactly these reasons. Females were observed too, but their centrality to the social group was very much underappreciated. It may not have been a coincidence that nearly all the scientists at the time were males too.[13]

While the early emphasis on male social chimpanzee behavior was due partly to observer bias, the lives of males are unquestionably more consumed with dominance status than those of females. An often-seen scenario is for two brothers, having grown up in a community together, to conspire toward a common goal. Sometimes their rivalry for status is just as important (see Richard III's overthrow of his hated brother Edward IV in William Shakespeare's play for a human example). The political relationship among males in a community is complex, sometimes but not always connected to kinship, and it changes over time. Chimpanzees live long, interesting lives, growing up alongside their allies and rivals, slowly reaching adolescence and adulthood, and struggling for a place in the dominance hierarchy. The results are not easily predicted. We long assumed that male cooperation in political contests was based on kinship. Since the 1990s, advances in field DNA sampling have allowed us to test that hypothesis. In some cases, it turned out to be incorrect or at least simplistic.[14] This added a new layer to research about the web of male political alliances, which appear to be as much about conspiring toward shared goals as about kin cooperation.

Why would a male chimpanzee care about his dominance rank? Achieving high rank takes years of effort, energy, stress, injury risk, and uncertain reward. Separate research by Michael Muehlenbein of Indiana University and Watts and by my former graduate student Martin Muller of the University of New Mexico and Richard Wrangham showed that high-ranking male chimpanzees incur increased levels of cortisols, a marker of stress. We know from Stanford University psychologist Robert Sapolsky's hallmark study of dominance and stress in baboons that increased parasite loads are also associated with high rank.[15]

The benefits of dominance rank have been pondered for decades, and the general consensus is that it must bring a reproductive benefit to be worth all the trouble. A link between mating success and rank is well documented for other mammals, from red deer to elephant seals. It's also been shown in primates. In those species living in one-male groups, in which one male tries to monopolize females, that male has nearly 100 percent paternity. This is seen in red howler monkeys, for example. Rates of over 80 percent are seen in baboons and patas monkeys.

Being an alpha or a high-status male primate does not, however, guarantee that he will mate with the most females or father the most babies. Some studies, such as Glenn Hausfater's early study of savanna baboons, have shown that dominant males have priority of mating access to females, especially when those females are ovulating. More recent studies by Duke University biologist Susan Alberts and her collaborators that have quantified paternity have upheld these findings. But there are also studies that show an apparent disconnect between dominance and mating success. University of California, San Diego, primatologist Shirley Strum found that newly immigrated males ended up at the bottom of the dominance hierarchy but nonetheless have high mating success. Alberts and colleagues showed that an unstable dominance hierarchy in particular allows the new male to take advantage of his "novel male" status. In the 1980s, Barbara Smuts found that female baboons preferred to associate with, and also mate with, males that were decidedly past prime, perhaps because they offered protection and support without putting her at risk of too much competitive aggression over sex.[16]

Because chimpanzees live in fission-fusion societies rather than cohesive social groups, males face a different set of issues in sorting out how their status may benefit their reproductive success. At least one study, by Rebecca Stumpf of the University of Illinois at Urbana-Champaign and Christophe Boesch, found the same female preference for lower-ranking males that Strum observed in baboons.[17] Most chimpanzee research has, however, found a more complex association between dominance rank and reproductive success. Fission-fusion society means the animals spend much of their lives dispersed in the forest—females in particular—so head-to-head dominance interactions may be fewer. The composition of the party changes hourly and may strongly influence the outcome of dominance contests. Low-ranking males usually attempt to mate with females only when out of sight of the alpha. If chimpanzees B and C can jointly challenge alpha male A, then all three must be together, perhaps abetted by the right association of hangers-on, for a dominance upheaval to happen. Even that is a simplified version of how dominance plays out in chimpanzee society. Usually the alpha rank is clear cut, but sometimes it is not, with two males sharing top status and eyeing each other warily. Occasionally, the community may be without a clear alpha, and dominance and mating may become a chaotic free-for-all that can continue for weeks or months.

Male dominance rank is often dependent on the social situation in which the participants find themselves. There are three-way and four-way social interactions, in which one animal's actions set up an instant chain reaction that extends to others in the vicinity. The same is true for alliances, which tip the balance away from more powerful, lone actors in favor of lower-ranking males who team up briefly. In my own field studies, there was always a single alpha male, but his power at a given moment was highly dependent on those around him. In Gombe in the 1990s, Goblin was a former alpha who had lost his high status and nearly died after a fight with other males. Only Jane Goodall's decision to provision him with water and antibiotic-spiked bananas ensured his survival. After a long recuperation from his wounds, Goblin eventually returned to the social arena. He never achieved alpha rank again, but performed brilliantly in the role of kingmaker, a high-ranking older male whose support was sought by up-and-comers.[18] This was a far better fate than that suffered by many alphas at the end of their reign. In Bwindi, the alpha was a powerful young male, Mboneire, who relied on a coterie of other males plus a female or two for the social support every alpha needs. He tended to avoid the chaotic challenges to his priority of access to ovulating females by taking a female with a swelling off on an extended consortship—a so-called safari—from which they would return only after her swelling period was over.

So why would a male chimpanzee strive to achieve high rank? The most obvious possibility, reproductive success, could not be directly studied until the past fifteen years, when DNA extraction from feces became practical. Before that time, we would examine the behavior records for any female who possessed a sexual swelling about 220 days before a given baby chimpanzee was born. The gestational range varies from about 205 to 240 days, depending on the study site; 220 days is the mean figure from Gombe. In a study of the Gombe chimpanzees, Emily Wroblewski of Stanford University and her colleagues examined paternity in relation to male dominance. They found, somewhat surprisingly, that most of the males at Gombe sired at least one offspring regardless of their dominance status. The number of mating opportunities was directly related to the

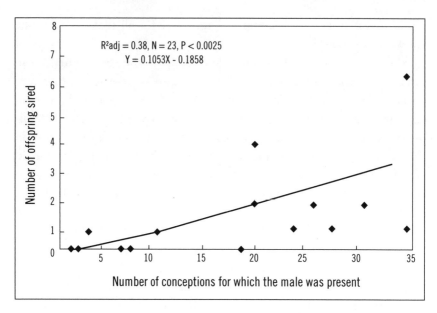

Gombe conceptions and reproductive success for males twelve years old and up.

number of offspring sired by males, which belies the time-honored idea that because chimpanzees are highly promiscuous, individual mating opportunities don't matter much. Most of the babies were fathered by teenage males. This doesn't necessarily mean young adult males were more attractive to females. It may be that high-ranking males achieve their rank by an early age and immediately begin to make use of it to sire babies. But the likelihood of fathering babies declined after age nineteen, even though the male's dominance rank continued to climb. Here we see the complexity of the issue: the relationship between dominance and reproductive success is not always straightforward. At Gombe, some low-ranking males fathered more babies than some high-ranking males, and in general they performed better than expected in the mating arena. This is not very different from the results of some of the baboon studies mentioned earlier.

Wroblewski and colleagues found, however, that male reproductive success did increase with increased dominance rank. The alpha male fathered nearly one-third of all offspring, which was 50 percent greater reproductive success than the next two most reproductively successful

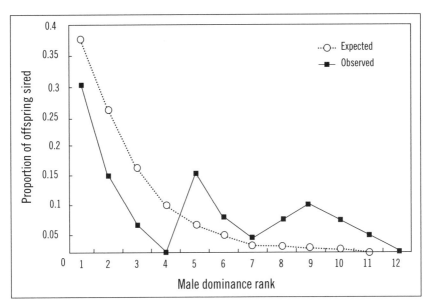

Male dominance rank and reproductive success, predicted by priority of access model and actually observed, Gombe.

males (who ranked second and fifth in dominance rank) combined. The overall number of adult and adolescent males varied over the course of the twenty-two years of the study period, but the alpha generally fathered more offspring than one would predict from chance. The effect was statistically significant, albeit not dramatic.[19] Among elephant seals and red deer, for example, the dominant male may father 80 percent of offspring in a given year. Now *that's* reproductive success.

At Gombe, it appears there is less reproductive payoff for males who achieve high rank than, for example, in Taï National Park. There, Boesch and his collaborators found a better fit for the priority-of-access model; the alpha male fathered fully half of all offspring whose paternity could be established. The more a high-ranking male competed with the alpha for mating opportunities, the lower the alpha's reproductive success. His paternity ranged from 38 percent, when five to nine other competing males were present in the community, to 67 percent, when only two or three male competitors were on the scene.[20] Whereas Wroblewski and her colleagues questioned whether the fission-fusion society of chimpanzees may limit the ability of an alpha to monopolize mating success—you can't mate-guard females who aren't anywhere near you—Boesch argued that the large party sizes and the large community size at Taï gave males access to many females and therefore the chance to father more offspring. It's important to remember, however, that female chimpanzees only mate when they have sexual swellings. Boesch and his colleagues found that, as you might expect, the more sexually swollen females there were at the same time, the more diluted paternity became. An alpha can have high reproductive success, but he can't monopolize all the females when there are many ready to mate and there's a gaggle of other eager males around. A notable exception to this pattern is at Bossou, where only a single adult male lived in the community and therefore fathered nearly all the babies.[21] The same paternity dilution effect has been well documented for other primates as well, such as baboons and long-tailed macaques.

Male chimpanzees do not have the probability of nearly 100 percent paternity that comes with alpha status in some social mammals. Nonetheless, in a long-lived animal like a chimpanzee, in which an alpha might hold power for several years and live more than forty years total, even a small edge in reproductive success per year could have a profound impact on the total offspring he sires during his life.

High rank may also be indirectly related to lifetime reproductive success. One day a few years ago, my University of Southern California colleague Caleb Finch and I were discussing social status and longevity. Finch, a renowned authority on the human life span, pointed me to studies of Western societies that have strongly linked high social and economic status to health, wealth, and longevity. A study had just been published in the *Journal of the American Medical Association* by S. Jay Olshansky of the University of Illinois at Chicago on the longevity of U.S. presidents.[22] Contrary to the popular perception that presidents age rapidly in office—at least based on the development of gray hair and wrinkles—they actually outlive other American men. If one removes the presidents who have been assassinated from the sample and considers only those who retired from office and lived out the rest of their natural life spans, U.S. presidents live long lives. In fact, the first eight presidents in the eighteenth and nineteenth centuries had an average life span of eighty years, which is about equal to the life expectancy of American women in the twenty-first century. Twenty-three of the thirty-four presidents who had died of natural causes had longer life spans than what would be predicted for them in the generation in which they lived; in some cases they lived dramatically longer. Presidents, especially early in American history, had better nutrition and better access to medical care than most people, having been drawn from the upper classes of society. But the results were still surprising.

Finch asked me if the same correlation would be true for chimpanzees; do high-ranking chimpanzees live longer? My first reaction was to say that they don't, because alpha males incur higher stress hormone levels and also a higher parasite load. We decided to investigate. My doctoral student Maureen McCarthy, Finch, and I compiled information on the tenure and life span of as many male chimpanzees as we could find in the published literature. We then compared the life spans of alpha males with all the other adult males who never became alpha. Because chimpanzees live long lives, we had to omit those who had been born before the various studies began and whose exact ages were unknown. We also omitted those who had not reached the age of fifteen—that being the age of the earliest known alpha male—at the time the data were published, and never-alphas who were older than fifteen when the study began.

We discovered that male chimpanzees who hold alpha rank for even a few months live significantly longer than those who are never alphas. The average life span for all males in the study who never reached the rank of alpha was 25.5 years. For alphas, it was 33.4. Alphas lost their top rank at about the same age that other males died; the alphas then went on to live an average of eight additional years. Eighteen percent of alphas outlived the very oldest male who was never alpha. Although the overall sample was small—there are only a handful of study sites with long-term data on alphas—the results were statistically significant.[23]

We know that reaching alpha status correlates with longer life spans, but we don't know that being dominant *causes* a longer life span. It could be the opposite; perhaps the most genetically and nutritionally healthy males are predisposed to reaching high rank. The finding paralleled that of the study of American presidents and also resembled the conclusions of long-term studies of life expectancy differences between upper- and lower-income classes in some Western industrialized societies. We suggested that there might be an ancient origin for the relationship between human life expectancy and social status. This could be due to direct effects: better nutrition or health care. It could also be due to indirect effects: being high ranking may carry psychological perks that promote long, healthy lives.

Signaling Submission

Two male chimpanzees approach one another from opposite directions on a forest trail. One is the alpha, the other an up-and-coming young adult male. As the distance between them shortens, the lower-ranking male stares at the alpha and issues a low-pitched pant grunt as they are about to pass. The two animals walk past one another without incident. But a few months later on the same trail, the same two chimpanzees have a very different interaction. This time, the lower-ranking male walks right past the alpha without pant grunting. It's like walking past your boss in an office corridor without saying good morning. The effect is immediate. The alpha wheels and charges at the disrespectful younger male and gives him his comeuppance. The alpha recognized the dismissive nongesture for exactly what it was—a challenge to his authority. In the coming weeks and months, the challenger will take on the alpha in ways large and small, and he will either be beaten back or instigate a dominance upheaval through his insubordination.

Vocal communication plays a key role in the lives of most higher animals, including the great apes. In studies of chimpanzee social status, we use the direction in which pant grunts are given to determine dominance. Chimpanzees are the only primates, and practically the only social mammal, in which one animal vocally announces its social status so overtly. Goodall and later Bygott, one of her earliest protégés, first documented pant grunting as a means of sorting out dominance relationships that went beyond body language or actual fighting.[24] The call is a powerful vocal signal, and it is often accompanied by some submissive body language—cowering a bit or, for a female, presenting a swelling.

The fact that the submissive chimpanzee doesn't always pant grunt is the fascinating part, because it means that he's made an assessment about the social interaction about to happen and decided whether to show obeisance. In one case in Budongo Forest, Nicholas Newton-Fisher reported that the alpha rank was held for a time by not one but an alliance of two males, who did not pant grunt to each another.[25] And though pant grunts are a routine part of chimpanzee social life, we don't understand the social factors promoting them very well. Because they're not obligatory, not only are they not always given when expected, there are times when both the presumed dominant and submissive chimpanzees give them to one another. This may indicate something further about mutual social assessment—each of the two apes may be uncertain about the perceived social status of the other. This context-dependent self-analysis is a window into how the chimpanzee mind works and a rare opportunity to see such cognitive assessment happening in the wild.

Marion Laporte and Klaus Zuberbühler of the University of St. Andrews examined the social variables that influenced when female chimpanzees in Budongo gave pant grunts to males. Since all females in a community are subordinate to all males, female pant grunting to males should always be unidirectional. They found that females adjusted their pant grunts depending on the context, indicating it is far from a rigid, ritualized call. They were more likely to pant grunt during positive social interactions than during aggressive encounters. In other words, females didn't necessarily use pant grunts to appease males or attempt to mitigate aggression from them. Females did not pant grunt at other males in the presence of the alpha male, who of course received profuse and obligatory pant grunting. It appeared that females "respected" the alpha's authority to the extent that they were not willing to show the same respect to the other dominant males. Laporte and Zuberbühler concluded that female chimpanzees use pant grunts as a means of figuring out a dominant male's mood that day.[26] Pant grunts may therefore be better understood as an important piece of a chimpanzee's social tool kit for building social relationships, instead of merely a direct consequence of the dominance hierarchy. In this sense, the authors note, pant grunts are not unlike human greetings.

Hormones and Dominance

It's a common assumption that testosterone is the basis for male machismo, aggression, and dominance. Research has shown that when two male tennis players compete, the outcome has a measurable effect on each man's testosterone levels: the loser's testosterone levels markedly drop, and the winner's modestly increase. Testosterone is widely marketed as a health supplement for older men, despite all evidence that testosterone injections do not actually boost energy, vigor, or sexual performance. Male fantasies of testosterone-induced rejuvenation aside, Alan Dixson, the dean of primate sexuality research at Victoria University in New Zealand, has described the actual role of testosterone in males as everything from motivation to the development of fetal and adolescent genitalia and physiology.[27] For instance, Pawel Fedurek and his collaborators found that testosterone appears to mediate the frequency of long-call pant hooting by male chimps at Kanyawara; changes over both day and month in testosterone levels were associated with pant-hooting rates.[28]

Ironically, much of the research on hormones and male behavior is far more easily done on humans than other animals. The researcher need only ask for a urine sample from the two tennis players. Wild nonhuman primates, which might provide a key baseline from which to interpret the results of human studies, don't volunteer their biological samples or pee or poop on command. But as researchers have devised clever ways of obtaining these samples, and new technologies have emerged to preserve samples and conduct analyses under field conditions, some fascinating information has come to light.

The original animal models for the influence of testosterone on male aggression were actually not mammals at all. University of California at Davis endocrinologist John Wingfield's studies of male birds have documented rises in testosterone levels during the spring mating season, just when males need to be in prime fighting mode to compete for nesting territories and females. Wingfield performed field experiments of male behavior by exposing the subjects to males implanted with testosterone; this led to an increase in their own testosterone levels. Seasonal spikes in testosterone were predicted given the need for males to show greater aggression and dominance when competing for mates. Wingfield called this the challenge hypothesis, and it has been the basis for many further studies of the impact of testosterone on mating behavior.[29] The causal connection in many bird species is clear; but in social mammals, including primates, it's less so. Sapolsky did groundbreaking work on the hormonal bases of social behavior among wild savannah baboons in East Africa. He found a nuanced connection between testosterone and behavior that varied according to social context. Dominant males become aggressive mainly when there are upheavals in the dominance hierarchy. These are periods when attempts to gain the upper hand suit them most. It is also during these phases of dominance instability that testosterone levels peak.[30]

High-ranking male chimpanzees are more aggressive than low-rankers nearly all the time, so the influence of circulating hormones must be different in this species. Muller and Wrangham studied the relationship between testosterone and aggression in male chimpanzees at Kanyawara in Kibale National Park, Uganda. Getting the testosterone samples was a challenge. Muller and his field assistants waited under the chimpanzees as they awoke at dawn in their leafy night nests high above him in the trees. Like all of us, they urinate when they wake up, and he collected samples by extending a long pole with a plastic bag attached underneath them, or by pipetting urine droplets off leaves. Using a filter paper system previously used for studying orangutan reproductive physiology, he recorded the contents of the urine.

Muller and Wrangham found that testosterone levels were highest in the morning, as is the case for other primates, including humans. These morning samples showed no association with dominance rank. Testosterone levels in samples collected in the afternoon did correlate with dominance; the alpha male had consistently higher testosterone levels than the other males. And, consistent with Wingfield's bird studies, male chimpanzees excreted the most urinary testosterone during periods when they were most aggressively competing over sexually receptive female chimpanzees. The effect was only observable with females who had previously given birth. Males did not have elevated testosterone levels in association with competition over females who had never given birth, even if they bore sexual swellings. Male chimpanzees tend to favor older females; perhaps those females are more exciting to them due to dominance or the right endocrine mix. High-ranking males were predictably more aggressive overall than low-ranking ones. The Muller and Wrangham study did not show cause and effect. It showed a testosterone relationship with aggression, but not necessarily with sex (since males with high testosterone didn't compete over certain sexually available females). It showed a difference between morning and afternoon sex hormone levels that might have affected male behavior that day. Studies of human hunter-gatherers have demonstrated much the same effect.[31]

Roman Wittig of the Max Planck Institute for Evolutionary Anthropology and his colleagues found that even a single aggressive interaction between two wild male chimpanzees can increase glucocorticoid levels.[32] Meanwhile, Muehlenbein and colleagues studied the testosterone-dominance relationship at Ngogo, also in Kibale National Park. Their results were consistent with those of Muller and Wrangham.[33] As in the Kanyawara study, the correlation held only during periods of social stability and not during rank upheavals. Both research groups attributed the uneasy fit in their testosterone-dominance results to the nature of chimpanzee social behavior. The fission-fusion mating system creates so much contextual unpredictability that challenges don't follow the bird pattern, making chimpanzees a poor model for Wingfield's challenge hypothesis.

Stressed Out

Social dominance in chimpanzee life also involves a major downside: stress. Sapolsky's pioneering study of social dominance in relation to sex hormones in male baboons was also a study of cortisols—hormones involved in stress and other body functions—and behavior. Early studies of stress and social context led primatologists to predict that in species with strong social hierarchies, such as chimpanzees and baboons, low-ranking animals would live with high levels of both stress and circulating stress hormones. The relationship actually applies in some species but notably not in others. Making matters even less clear, in those species in which social dominance (or, more properly, submission) and cortisols are associated, that association may break down at times of dominance upheavals, exactly when you might predict it to be at its strongest. Perhaps the social upheaval itself is so stressful to all involved that both high- and low-ranking members of the hierarchy suffer a stress response. It may be that stress and high levels of stress hormones are just the cost of doing business as a social mammal.[34]

Muller and Wrangham argued that, in chimpanzees, the high energetic costs of aggression—which so often involve prolonged and dramatic displays and sometimes fights with other males—may be a factor in the production of stress hormones. The alpha often wakes up, climbs down from his sleeping tree, and starts his day with a long and energetic charging display around the forest. Muller's work on wild male chimpanzee cortisol indicated an interrelationship of aggression, dominance, and stress. First, he found a nearly linear relationship between a male's dominance rank and the frequency of his aggression. Second, levels of cortisols were correlated with aggression in the afternoon, but not in the morning (Muller and Susan Lipson had already shown that morning hormone levels were highly variable). Because morning urine samples came after many hours of sleep, males presumably woke up relatively unstressed. That stress built through the day as they experienced the small and large stressors of social life. Muller and Wrangham did not find, however, that high-ranking males incurred high levels of cortisols due to increased stress.[35]

The point is that all of the stress is worth the trouble to a male chimpanzee if in the end he fathers more babies. As we saw earlier, this appears to be the case, but not as markedly as one might think. A male chimpanzee must endure the trials and tribulations of striving for dominance—stress, parasites, and injury risk—because the result is potential alpha status and the enhanced reproductive success that comes with it. But if that reproductive gain is not always forthcoming, then the argument for dominance becomes circular. Primatologist Thelma Rowell of the University of California, Berkeley, said that we should call the dominance hierarchy a submission hierarchy because group members tend to learn their place and stay in it.[36] Challenges to higher-ranking groupmates are relatively rare and always risky. Even if successful, the social upheaval that may result can bring stress to you and your kin and, when the chips fall, the new social order may not be to your liking.

Female Rank

The quest for dominance is not limited to males. The strong hierarchies that characterize the interactions of males do not seem to occur among females, but this doesn't mean it's not important in their lives. Primatologist Carson Murray and her collaborators studied female chimpanzee dominance and reported female-to-female aggression interactions and pant grunts over a two-year period. They were able to establish a linear hierarchy based on a much smaller sample than one would expect for males.[37] Steffen Foerster and Anne Pusey of Duke University and their colleagues described categories and trajectories of rank for the females of

Gombe's Kasekela community. They found that females tend to wait patiently for opportunities to increase their rank, which is very different from the male strategy of actively challenging everyone above them.[38] Wrangham and colleagues reported much the same for females in Kanyawara in Kibale National Park, Uganda.[39] In the Sonso community in Budongo Forest, primatologist Katherine Fawcett did not believe females in the community could be consistently sorted into broad dominance categories.[40] Wittig and Boesch investigated female dominance relationships among the chimpanzees of Taï National Park in Ivory Coast. Pant grunts were less useful for parsing female dominance and submission than they were for parsing that of males. In a third of female-female encounters, pant grunts either were not given or were mutually given. Overall, female greetings happened at one-sixteenth the rate as among males during the same time period. This is a consequence of fission-fusion society; females just don't socialize as often as males do. However, using the much smaller sample the females provided, researchers discerned a linear hierarchy that sorted females into well-defined categories. Squabbles over food were actually the best way to establish female dominance in the Taï study. Competition for access to shared meat—a passingly small fraction of the diet—characterized most of the female fights in which something that could be monopolized was at stake.[41]

Wittig and Boesch found that the more females there were in the feeding party, the higher the conflict rate. When they examined how the linear ranking mattered in females' lives, they found that dominance rank was related to age—older females were dominant over younger ones. This is different from the male pattern, in which alphas are usually eighteen-year-old to twentysomething males, and males in their thirties and over are typically either past alphas or never-alphas. But the researchers also found that female dominance was based primarily on the outcomes of food squabbles, with age only a secondary factor. The relationship among females in a community is not close, and sometimes not cordial. They use aggression to take food for themselves or keep other females from taking it. Over time, the accumulated outcomes of such contests determine the female hierarchy, with older females better able to dominate food sources. Since meat is one key food that can be monopolized, and since males have favorite females with whom they prefer to share

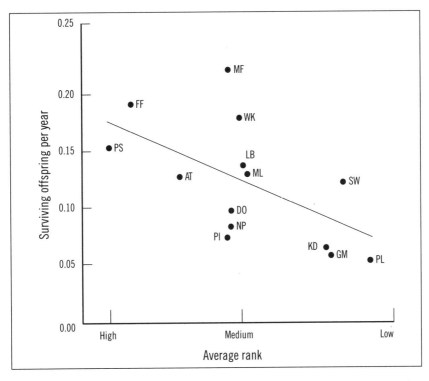

Number of surviving offspring per Gombe female in relation to dominance rank.

meat, meat-sharing episodes can help to determine the rank categories of females.

Even though dominance seems to define females less so than males, dominant females do reap important benefits. As with males, we have to ask what benefits are worth the trouble a female goes to in striving to assert herself over all other females. Conceiving and producing healthy infants and then rearing them successfully is what a female chimpanzee's life is all about. Pusey and her coauthors analyzed long-term data from Gombe Stream National Park and found an association between a female's dominance rank and the number of babies she had, as well as how likely those babies were to survive infancy and how rapidly they grew up.[42]

Pusey and colleagues point out that we don't understand well the social process through which female chimpanzees achieve high rank. Using the direction of pant grunts given and received, they assigned females to the broad categories of high, middle, and low ranking. Older females tended to be more dominant, as in other studies. High-ranking females reproduced more rapidly, although this finding only held true by omitting Gigi, an older, high-ranking female who had never given birth and was speculated to be infertile, from the analysis.[43] The daughters of high-ranking females reached puberty up to four years earlier than those of low-rankers. Given the long interval between births of baby chimpanzees, that could amount to one extra offspring produced in each highly ranked female's life span. This is a huge reproductive edge that may be related to social dominance. We know from work by Boesch at Taï that dominant females invest more time in rearing sons than daughters, improving the rate of male infant survival.[44] Male dominance and competition among chimps don't seem to be based on body size—many high-ranking male chimpanzees are small bodied, unlike gorillas and many other species. The preferential treatment of sons by high-ranking mothers must have a cause and a consequence. High-ranking females also lived slightly longer. Since female chimpanzees do not experience menopause, this may result in more offspring produced.

Much of the dominance effect among females likely comes down to food. High-ranking females at Gombe occupy core areas in the Kasekela community, which may provide them a more reliable and productive food resource base. Some females from certain powerful lineages at Gombe don't even emigrate at sexual maturity, perhaps because their core area is ideal foraging habitat over which they maintain prime access. Murray and colleagues studied the effect of female dominance on the use of forest space at Gombe and found that newly immigrated females took up residence in areas of the forest away from those used by the resident high-ranking females. These new and low-ranking females wandered; they were less likely to travel in any single core part of the community's range. Dominant females used smaller core areas, suggesting that they needed to expend less energy searching for good food.[45] The implication is that dominant females monopolize the best food areas, which they know well from many years foraging there. This gives them an edge in finding the best food, especially during lean seasons, and this edge translates into greater reproductive success and infant survival.

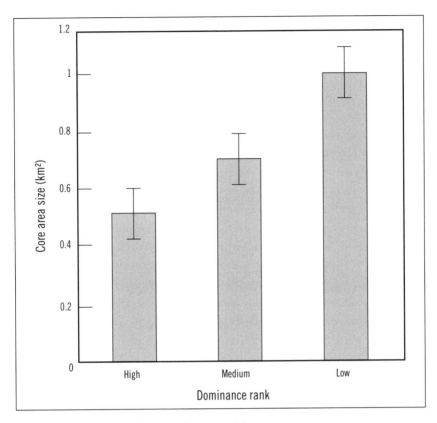

Female dominance in relation to their use of forest space.

Are Dominant Chimpanzees Born or Made?

By now, I hope I've made two points clear. First, we don't fully understand why a male chimpanzee wants to be the alpha or even high ranking. Reproductive success must be associated with dominance, directly or indirectly, and we have evidence that this is the case. Second, we also don't know exactly how chimpanzees become high ranking, especially in the case of females, whose dominance interactions are too few and far between to compile an ideal data set supporting or refuting various theories. The question that underlies all of this is, are dominant chimpanzees born or made? Infant males spend years growing up. They learn their roles in the community, and, as adolescence turns to adulthood, they begin to challenge the higher powers in the hierarchy. For the communities for which we have good long-term genealogical information, males who reach alpha status and hold on to it for a few years have tended to descend from lineages of high-ranking chimpanzees. Four related Gombe males—Figan, Freud, Frodo, and Ferdinand—held alpha status for a combined total of more than twenty-two years, nearly half the length of time that chimpanzees at Gombe have been under observation.[46] Add in Goblin, who was the son of high-ranking female Melissa, and about two-thirds of recorded Gombe chimpanzee history has been ruled by two powerful matrilines. Although there are alphas who rise to high rank without the key connections of family, most come from lineages that have produced prior alphas. Long years of strategizing, politicking, and storing favors owed by potential allies gets these males to the top, assisted by an inordinate amount of pluck.

It's impossible at present to tease out the effects of learning and experience from the influence of genes when assessing the underlying bases of high or low rank. We have detailed studies of male political strategies and tactics and how they are aided by family and ally connections. That is where most researchers lay the credit for achieving high rank. But we cannot eliminate the possibility that high-ranking animals are genetically predisposed to dominate their peers. If a tendency toward high or low rank is genetically inherited, we would expect very much the same pattern that actually occurs. In addition to asking whether high rank leads to greater reproductive success, we would then have to ask if it might not be the other way around. Perhaps a lineage's reproductive success promulgates babies that are themselves predisposed to eventually achieve high rank. Since behavioral indicators of rank are less obvious in females, one could argue that they provide a better argument for a role of genes than males do.

Most male chimpanzees seem so driven to ascend the social scale that something good must be waiting for them at the top. And while they're endeavoring to do that, a great deal of fascinating political machinations take place. Some animal researchers claim that chimpanzees are little different from other mammals in their social strategizing. Few chimpanzee researchers would agree. The complexity of chimpanzee society is not unmatched; elephants and dolphins live in social systems that may be every bit as layered and contextual as that of chimpanzees. But chimpanzees (and the closely related bonobo) take political behavior to a level seen otherwise only in humans. Some of this political sophistication may be due as much to the random course of evolution as to brain function. If you live in the ocean and have no hands, like dolphins, then the context and the power of social manipulation may be quite different. But great apes, and in particular the genus *Pan*, have taken the suite of brain size and associated traits to the highest level known in the nonhuman animal world.

CHAPTER 4

War for Peace

CHIMPANZEES USE AGGRESSION in ways that repulse us when we see it in our own species. My University of Southern California colleague Christopher Boehm estimated that the rate of nonlethal violence among wild chimpanzees is greater than that of most human societies.[1] A separate study by Richard Wrangham and his coauthors found a similar "murder" rate between chimpanzees and traditional human hunter-gatherer societies, and a much higher rate of nonlethal aggression by chimpanzees.[2] Chimpanzees are the only primates other than us who routinely kill one another in the name of territory and resources. As in human societies, the killers are virtually always males, as Wrangham pointed out in his book *Demonic Males*.[3] Chimpanzees lack the weapons we associate with efficient killers; they have hands and fingernails, not paws and claws. Their canine teeth, while impressive, are no match for those of a carnivore. And yet they carry out grisly attacks on members of their own and especially neighboring communities. Males sexually coerce females. And both males and females are known to commit infanticide.

Chimpanzees are not killing machines; 99 percent of their lives are spent in peace. Of course the same could be said about us. The potential for violent behavior is within each of us, but it surfaces only rarely, or never at all. And just as humans have myriad ways to defuse disputes before they reach a stage at which violence seems a feasible option, chimpanzees have many fail-safes that prevent lethal aggression from taking place. After minor squabbles they reconcile, and the ways in which they restore social harmony are as interesting and important as the violence that gets all the attention from scientists and the media.

Natural-Born Killers?

Like every other mammal on the planet, chimpanzees have the capacity to inflict physical harm on one another. It's harder to take a utilitarian approach to understanding violence in chimpanzees than it is in lower mammals. Chimpanzees who injure or kill one another are not immoral. They are amoral; their violence is a means to reach an end. We don't get angry at lions for attacking each other or for killing zebras; that's what lions do. We tend to view great apes in a different light because of their close evolutionary connection to us. An entire wing of animal behavior research is founded on the idea that the roots of human morality may be found in the premoral behavior of nonhuman primates, with chimpanzees serving as a prime animal model. Most researchers have concluded that "might makes right" when it comes to chimpanzees' treatment of one another, but that hasn't stopped anthropologists from citing chimpanzee aggression as a potential example of how punitive violence may have its cultural origins in our own species.

There is a school of thought—a poorly informed one—that holds that chimpanzee aggression is somehow the product of human interference in their behavior. This charge was first raised shortly after Jane Goodall first observed the males of one Gombe community seeking out and attacking the males of the neighboring community. Since these two groups had recently split from one another, males were ganging up to kill their former comrades. Some anthropologists argued that the violence was precipitated by the presence of human researchers, or by humans provisioning chimps with bananas, or by humans altering the habitat, or perhaps even by the habituation process itself. The argument gained traction especially among scholars who view human societies as egalitarian and peaceful by nature.[4] They argue that intergroup violence is the product of outside forces such as Western contact. We then learned that such killings between males of adjacent chimp communities happen at nearly every research site where the apes have been observed. There was even a report of wild, unhabituated male chimpanzees attacking a group of captive chimpanzees in a facility in Senegal that happened to be inside the wild males' territory.[5]

The stakes are high. If male chimpanzee violence is adaptive rather than pathological, we might infer that the same is true for humans. Male chimpanzees use violence to achieve resource-related goals—food and sex—by eliminating rivals for both. The argument that extreme violence is an aberration dissolved with more and more field observations of chimpanzee violence. For most of the past three decades there has been a consensus that violence is a normal, adaptive behavior among chimpanzees. A 1991 book by anthropologist Margaret Power attempted to resurrect the idea of chimpanzees as peaceful by nature and violent only when their social behavior is disturbed by human influence. Anthropologists Robert Sussman and Joshua Marshack of Washington University, Saint Louis, made a similar claim. These authors claim that human-caused habitat disturbances, combined with small forest size and provisioning, can produce increased, even lethal, aggression.[6] The skeptics' case ignores the voluminous data on chimpanzee violence. The most frequently violent chimpanzees that we know of, at Ngogo in Uganda, live in one of the most pristine habitats in which chimpanzees have been studied.

Michael Wilson of the University of Minnesota and his colleagues recently analyzed the pattern of chimpanzee violence that has been compiled over a half century of field research. They considered potential predictors of violence that are human caused: habitat disturbance, provisioning, and size of the forest. Wilson and colleagues found that none of these factors predicted which chimpanzee populations displayed the most violence. Instead, the best predictors of violence were adaptive factors. Violent attacks are more likely when there is an imbalance in the number of males in the parties of two adjacent communities. Attacks are also predictable from male demography; attackers are normally sexually mature males, so the more males in a community, the more violence we see.[7]

It might have been acceptable in the 1970s or 1980s to be skeptical about the adaptive nature of chimpanzee violence, but with the accumulated observations of wild chimpanzees since, it cannot be written off as "unnatural." In this chapter I'll examine several modes of chimpanzee violence. All are "natural" in the sense that they occur routinely in the wild under a wide range of environmental conditions. In the soul of their biology, chimpanzees possess the potential for behaviors that we consider immoral when we see them in ourselves. When they occur in chimps, they are simply strategic ways to achieve life goals. Deciding whether there is an evolutionary link between chimpanzee aggression and human violence is an important topic, but not one that should influence the interpretation of chimpanzee behavior.

Murder Most Foul: Violence within the Community

In penal codes in societies around the world, murder is most often defined as the taking of another's life without justification. It is the killing of another with malice. This is of course a definition made for and by humans. Applying it to a nonhuman animal, including our closest kin, is highly problematic. We don't call chimpanzee violence a crime, and if there is any punishment for violence it comes in the form of retribution meted out either in the moment or sometime later. This leads to an uneasy comparison with human murder. Even if the checks and balances that control bad behavior in our own societies are purely cultural, is the motivation for violent crimes ever comparable? When a cabal of male chimpanzees seeks out and gangs up on another male in their community, a biological justification can always be found. He's a genetic rival, a competitor for food resources, or a political rival. In human societies, none of these justifications matter in deciding the punishment, because our moral code says it's wrong to take another life in nearly every circumstance other than self-defense. When a male chimpanzee takes a rival out of the game, he creates mating opportunities for himself and perhaps increases his chance of becoming alpha someday. Wiping out the males of a neighboring community may bring added territory, new fruit trees, and new females into play for the killers.

The evolutionary rationale for chimpanzee violence is clear, in terms of reproductive success. The past decade and a half has produced exciting new observations of violent behavior within and between communities. Brutal attacks carried out by males on other males in the same community have been observed at three chimpanzee study sites: Mahale Mountains National Park in Tanzania, Ngogo (within Kibale National Park) in Uganda, and Budongo Forest Reserve in Uganda. Killings within the same community are rarer than killings between communities. Even nonlethal attacks require some explanation, because the motivations for them may differ from intercommunity "warfare." When several males seek out and gang up on a lone rival of the same community, the killers are removing a potential ally in intergroup territorial conflict. Sometimes the victim is an alpha, and the goal of the cabal is clearly to overthrow him. Other incidents have involved a younger male who was prematurely asserting himself and older males who decided to put an end to that, and to him. In the Budongo case, the victim was a low-ranking male who did not pant grunt to more dominant males. Bad move—a group of several adults got together and brutally killed him.[8] At Mahale the victim was a recently deposed alpha who was found dead (the attack was inferred from his wounds).[9] Goblin at Gombe might well have suffered a similar fate—he misjudged his influence by trying to return to the community after being deposed—had the decision not been made to revive him with medication after he was severely injured.[10]

A killing in 2002 at Ngogo offers a window into the violence that occurs in communities. David Watts reported on the death of an up-and-coming young adult male, although the attack was not directly related to a struggle for alpha rank. Many of the adult males in the large Ngogo community had socialized with the victim; he had been part of a patrol of fourteen adult males two weeks earlier. A group of these males launched a brutal attack on the victim; at least seven of them inflicted deep bites while holding him to the ground over several minutes, and they left him mortally wounded. What was likely his carcass was found in the same area days later.[11]

How did this seemingly good citizen of the Ngogo community become an innocent victim? Watts points out that the victim had been low ranking several years earlier but was rising rapidly in status. In doing so he had violated the rules of chimpanzee social etiquette. He was less apt to be found in the company of the other males than one would expect, and he rarely groomed them. Perhaps most ominous, he did not form alliances with other males—these would have served him well when a subset of the males decided to attack. This lack of socializing is typical of low-ranking males, but for a social climber with big aspirations, it's a major faux pas. Imagine a young member of the U.S. Congress who had blind ambition but a lack of social common sense and so didn't create coalitions for himself as he climbed the congressional ladder. It would likely be his eventual undoing. The Ngogo victim had clearly become threatening to the higher-ranking males by not employing the social skills to defuse aggression from them.

It's not clear whether males who carry out intracommunity attacks are attempting to mortally wound or kill victims. It could be that the Ngogo males were only trying to punish the victim for his lack of social graces. Perhaps the attackers got carried away and inflicted wounds more serious than intended. Such retribution motives have been ascribed to intracommunity attacks at Mahale.[12] Watts described the Ngogo attack as "less severe" than those that occur between males of different communities.[13] This possibility is plausible if only because mortal woundings in intracommunity scuffles are rare—just a handful in a half century of field research in forests across Africa.

These examples of intracommunity violence all involved males attacking males. But males also attack adult females in their own communities, though with a very different agenda and outcome. Over the past decade, observations have mounted of male attacks on females, which have the apparent intent of intimidating them into mating. Aggression being a routine aspect of male chimpanzee behavior, this should surprise no one, but the intent to sexually coerce females has revealed yet another parallel with ugly aspects of human male behavior. It's been reported at low rates for many years, from sites across Africa: Mahale, Budongo, Ngogo, Kanyawara, and Taï. The incidents were too few and far between to identify an explanatory pattern. But that has changed.

Martin Muller and his collaborators analyzed a decade's worth of evidence of male sexual coercion at Kanyawara and published it in a volume on the topic in animals, including humans, coedited by Wrangham. They found that males display aggression toward females at almost precisely the same rate as they do toward other males. This doesn't necessarily mean physical violence. Two-thirds of aggressive incidents occurred without physical contact at all: bluff charges, displays, and the like. Adult males were far more likely to be aggressive toward females with sexual swellings, who therefore might be ovulating. Adolescent males, on the other hand, were most likely to be aggressive to noncycling females, suggesting that their coercion was not related to sex.[14]

Muller and his colleagues weighed different explanations for the level of male violence toward reproductively active females. Perhaps coercing females with impunity is yet another way that males display their dominance status. Or maybe they use females as pawns to signal their own status to rival males. This has been suggested for men in some traditional human societies. It's also possible, the researchers argue, that males use aggression against females as a way of "policing" female-female competition. Such policing may limit the extent of antagonism by resident females against new or potential immigrants, whom males want to encourage to stay in the community.

The only proposed function for chronic male aggression toward females that made sense to the researchers was sexual coercion. Males mainly target fecund females, who suffer injury and stress from male attacks. Furthermore, males that are sexually coercive reap a reward of increased mating success. Female chimpanzees are promiscuous and distribute their matings to confuse paternity. Male attempts to control female sexuality may be intended to punish females for actively seeking multiple mates.

Another recent analysis provided the potential rationale for males to coerce females. Joseph Feldblum and his coauthors analyzed nearly two decades of observations from Gombe Stream National Park and found that Gombe males are, like males elsewhere, frequently aggressive toward females. Furthermore, male aggression was especially frequent during females' swollen periods. And aggression by high-ranking males toward females when they were not swollen predicted males' likelihood of fathering babies with those females. In other words, high-ranking males use aggression and intimidation to boost the number of infants they can father.[15]

Male sexual coercion as a reproductive strategy seems effective in a species with promiscuous females that advertise their sexual status and live in fluid social groups in which it's hard for males to monopolize mates. Gombe and Kanyawara are the only study sites at which this finding has been made or suggested. Rebecca Stumpf and Christophe Boesch found that direct male coercion of females in Taï National Park was rare, and it did not seem to function as a mechanism by which males controlled female matings. Male aggression at Taï did not correspond at all to female sexual availability, unlike in the Gombe study.[16] In fact, the two studies produced such opposite results with respect to nasty male mating strategies that only when further studies from other field research sites are done will we be able to say with confidence that male chimpanzees everywhere use sexual coercion to improve their reproductive success.

You shouldn't get the idea from the foregoing that male chimpanzee nastiness elicits no response from females. It's true that female chimpanzees do not habitually form powerful coalitions to intimidate males and repel their advances (in the way that female bonobos do). But female chimpanzees do sometimes repel male aggression. In captivity, where the makeup of the community is artificial and may contain a powerful female or two who can tip the balance against males in a way that rarely happens in the wild, we see more female empowerment. In the Sonso community in Budongo Forest Reserve, females sometimes form coalitions to retaliate against aggressive males. Nicholas Newton-Fisher found that when attacked by males, females sometimes formed all-female coalitions of from two to six members, one of whom was always the victim of male aggression in the moments before the retaliatory coalition was created. The female coalition then counterattacked the marauding male, usually with screams and barks but occasionally with physical assaults. Males fled when attacked. High-ranking females invariably came to the aid of lower-ranking ones.[17] Such female alliances, so common in the closely related bonobo, are quite rare among chimpanzees.

Dogs of War: Violence between Communities

Wild male chimpanzees have, to say the least, hostile relations with males of neighboring communities. They're not alone among mammals or among primates in this respect. Wolves stage raids into the territories of other packs, injuring or killing rival neighbors. Male lions form coalitions to attack other prides, often resulting in the deaths of cubs. Male spider monkeys have been reported to conduct violent raids against other groups.[18] But lethal raids by chimpanzees have attracted by far the most attention because of the striking parallels to traditional human intergroup violence.

Chimpanzee intercommunity conflict doesn't exactly fit the label of warfare, which implies entire communities doing battle against each other. Traditional human hunter-gatherer societies engage in intergroup lethal combat, but homicide of one individual victim is far more common. Chimpanzee intercommunity conflicts are really raids. A party from one community attacks one or a few individuals from an adjacent community, usually in the overlap zone of their territorial boundaries. Such attacks may be carried out strategically when the attackers detect an imbalance of power. Ten chimpanzees rarely engage in a battle with ten or more enemies. First, few communities that we've studied are large enough for such large-scale battles (the huge community at Ngogo being a notable exception). Second, large-scale encounters are very dangerous for individuals, compared to attacks mounted against a much smaller number of rivals. So male chimpanzees monitor their territorial boundaries, picking and choosing their battles based on their perception of when a critical mass in their ranks can successfully challenge neighbors. We've observed this basic pattern for forty years now. In recent years, experimental field research and continued observations of intergroup violence have greatly enriched our understanding of how and why intercommunity conflicts happen.

Coalitionary aggression is a prime feature of chimpanzee attacks. Each member no doubt assesses his costs and benefits of participation. The odds of reward in the form of expanded resource access—more territory means more food and more mates—are weighed against the risk of injury or death. A recent study by Liran Samuni of the Max Planck Institute for Evolutionary Anthropology showed that oxytocin levels peak before and during intergroup conflicts, much as is seen in humans.[19] Wrangham points out that chimpanzee lethal raids are unusual in that they don't arise or escalate from larger conflicts. It's somewhat the opposite; when carried out over weeks or months, as has been observed at Gombe and Mahale, the males of one community may kill their rivals in a neighboring community.[20] This gives the appearance of a systematic extermination of the neighbors. The attacking community then takes the spoils of the attacks: new territory, new food trees, and new females. Neighboring chimpanzee communities may have access to very different levels of food abundance, so group territoriality may also have a positive impact on individual nutrition.

Violent intercommunity conflicts have been observed or strongly inferred in fifteen of eighteen well-studied chimpanzee communities: Gombe (Kasekela, Kahama, and Mitumba), Mahale (M and K), Kibale (Kanyawara, Ngogo, and Kanyantale), Kyambura, Taï (East, North, and Middle), Goualougo (Moto), Kalinzu (M), and Fongoli.[21] These conflicts may be carried out over long periods of time, and they can lead to the destruction of communities. During the 1970s, M group males at

Mahale likely killed as many as five adult males of neighboring K group, although the killings were not observed. Yukio Takahata of Kwansei Gakulin University recently reexamined the evidence and argues that at least some of the K group males were likely the victims of intercommunity violence.[22] During the mid-1970s, Gombe males of the Kasekela community attacked and killed at least five and likely more males of the southern Kahama community from which they had recently split. In both these cases, there was an imbalance of power in the number of males likely to be found in parties from the warring sides, which may be the explanation for the hostilities' escalating to the point of lethal raids by one side. As the imbalance of party size grew through the killing of now less numerous rivals, the incentive for further attacks by the larger group only increased.

Intercommunity violence typically begins when a party of males from one community sets off toward the territorial boundary they share with an adjacent community. This could be an apparent patrol of the territorial border or simply a foraging trip to fruit trees growing near the border, which turns into an intercommunity event when chimps from the adjacent community are encountered. Patrols happen unpredictably, often abruptly, and without a signal visible to a human observer. They happen at irregular intervals, from a few days to many weeks, and their frequency may be a function of the current relationship with a neighboring community. Goodall and Wrangham point out that during the infamous community split and subsequent warfare in Gombe in the 1970s, border patrols were frequent and were directed primarily in the direction of the rival community.[23]

A patrol begins when a group of males breaks off from some other activity and makes a beeline toward parts unknown. The males travel with increasing caution as they approach the territorial border. The border might be a riverbank or a forest break, or it may simply be a spot in the forest where one leaves the exclusive territory of the community and enters a "no-chimp" overlap zone. The males appear to be on edge, freezing at distant sounds to listen intently before continuing. They begin to show intense interest in objects that might be evidence of the enemy, stopping to examine and sniff stick tools, leaf wadges, nests, and feces.[24] Sometimes the males only skirt the border, checking for signs of the enemy before turning back into their own land. But the patrol may penetrate farther, making a deep incursion into enemy territory. The tension is palpable as the males continue for hundreds of meters before turning back. On most of the patrols I accompanied at Gombe in the 1990s, such a deep incursion ended with the males freezing when they heard distant calls from the enemy community, then wheeling around and racing back into their home range, whereupon they hooted and displayed as though venting the emotional tension of their mission.

We have to ask why all the males so eagerly head off for hours on a patrol. The incentive is obvious for higher-ranking males; the booty may be new females and food, which they can control. For low-ranking adult males and females (who also accompany patrols), the potential rewards are less, but the risks are still substantial. Males clearly incorporate territorial defense into their weekly foraging, but females have much less incentive to do so. Lucy Bates and Richard Byrne of the University of St. Andrews found that Budongo male chimpanzees tend to employ an "always-forward" foraging strategy, in which they resume travel after a halt in the same direction as before the break. This may be designed to take males in the direction of territorial borders in the course of daily wanderings. Lactating and pregnant females did not display this pattern, instead tending to circle back toward previously explored feeding areas.[25] Females should be expected to focus their energies on obtaining the highest-quality food for themselves, their infants, and their unborn fetuses. Males are far less constrained by the need for constant access to optimal resources. Females are less likely to be involved in territorial patrols or border disputes; patrolling is a mainly male behavior.

Watts and John Mitani studied border patrols at Ngogo. Patrols were observed on average once every ten days, about twice as often as at most other chimpanzee study sites. This may be due to the large number of males in the Ngogo community; more patrol members means lower individual risk of injury during an encounter for each male. Or perhaps there were simply more males on hand to start a patrol whenever the urge arose. A minority of males, somewhat more than a third, took part in every one of the patrols at Ngogo.[26] This low fraction is perhaps explained by the number of males overall in that community. This is in contrast to Taï, where three-quarters of the males took part in the average patrol.[27]

Patrols occur at or near territorial borders, although they often start deep within the community's territory. Territorial ranges expand and contract over many years, and the length and pattern of patrols appear to shift as well. When males are few, they tend to avoid patrolling in areas where they may be outnumbered by males of the neighboring community. Most often, patrolling males don't encounter the enemy at all. When they do, it's usually just hearing and responding to distant calls. Sometimes these distant calls are replied to, sometimes they elicit further approach or retreat, and sometimes they cause outright fright and flight by the listening patrol. Of fifty-two patrols studied by Watts and Mitani, only five led to physical attacks on neighbors.[28] Still, an attack rate of 10 percent means there is a nontrivial chance that someone will be injured or killed during a given patrol.

Watts and Mitani found that male participation in border patrols was related to their mating success, to their participation in hunting parties, and to their hunting success. This seems surprising until you consider that males simply tended to spend their time with the same male allies a great deal, so male-bonded activities such as hunting and patrolling involved most of the same males. The males formed coalitions within the community, and these same alliances became part and parcel of intercommunity defense and raiding. Despite the fluid and fragmented nature of the fission-fusion society—or perhaps because of it—males were more likely to risk life and limb when in the company of males with whom they had a strong bond. Marissa Sobolewski of the University of Rochester and her colleagues showed that the Ngogo males display an increase in testosterone while patrolling, as you might expect when the risk of a violent encounter with the enemy is greatest.[29]

Ngogo is perhaps a special case because of the outsize number of adult males on hand to patrol. A different dynamic applies at Taï, because females are much more involved in patrols. Boesch and his colleagues believe that chimpanzees in West Africa may have a fundamentally different set of intercommunity relations from those of the more thoroughly studied populations in East Africa. For most of the three-plus decades of observation of the Taï communities, lethal aggression has been extremely rare. In fact, no intercommunity killings were observed in the first twenty-three years of research there. The death rate at Ngogo, by contrast, has averaged about two victims per year.[30]

Taï males use intercommunity encounters as a means of kidnapping and keeping, at least temporarily, sexually available females. As in other chimpanzee studies, young adult females willingly visit neighboring communities, sometimes settling into them long term. This is in contrast to the rarity of such visits or transfers by older females that have already given birth. Unlike other chimpanzees, females at Taï tend to be quite involved in intercommunity encounters, lending support to the males. It is more common at Taï than elsewhere for adult females to be separated from their home community by enemy males during encounters, yet not subjected to severe violence by them. More than a third of the time, males and females from opposing communities have sex during such encounters.[31] This is unthinkable at most other chimpanzee sites. Females at Gombe are often the victims of severe or lethal aggression during intercommunity interactions (in 75 percent of encounters in the early years of research, and about half the time more recently). This violence at Gombe is directed at nonswollen females and at fertile ones, and it never includes intercommunity sex. The results at other sites such as Mahale are more mixed but decidedly less affiliative and more violent than at Taï. It's not entirely clear what's going on here. It's been suggested that females at Taï who mate with the "enemy" actually belong to both adjacent communities that share overlapping territories. Taï females are more gregarious overall, and perhaps therefore more sexually available than those in other sites in Africa.[32]

The First War

The foregoing discussion of the nature of intercommunity encounters was unthinkable in the early 1970s. Goodall had been conducting her Gombe study for more than a decade, and other scientists and students had joined her research effort to untangle the complicated web of behavior she had observed. Down the coast of Lake Tanganyika, the Japanese team led by Toshisada Nishida was solving the puzzle of community structure and aspects of behavioral ecology. A portrait of chimpanzee life had emerged that fascinated both the scientific community and the public. Then, in 1972, the main Kasekela study community at Gombe began to split in two.

The gradual split meant that a new community, the Kahama chimpanzees, now lived just to the south of the Kasekela chimps. It's not known what precipitated this fission. A speculative consensus exists that the community had grown large and was in the early stages of dividing when

Goodall arrived in 1960 and began provisioning the chimpanzees with bananas to facilitate habituation. This bounty may have kept the community together an extra several years.[33] A recent analysis by Feldblum and his collaborators suggested that a dominance struggle within the original large community may have led to the fission among the males; whether banana feeding played a role in that is unclear.[34] Whatever the reason, the original nineteen adult and adolescent males that were using the banana feeding station in the mid-1960s had separated into two factions by the early 1970s. Goodall began to speak of a northern subgroup and a southern subgroup, with Kakombe Valley (where the feeding station was located) at the center between the two.

By the start of the 1970s, the males that tended to spend their time in the northern sector rarely traveled southward or mingled with their former group mates, and vice versa. The northern and southern factions contained, respectively, eight and six adult males. Although the northern subgroup was larger and contained the former alpha Humphrey, the southern subgroup boasted a coalition of two high-ranking males—Charlie and Hugh—who together were dominant over everyone else, including Humphrey. When the two factions of males met one another in the forest—which increasingly happened only when a party from one subgroup traveled into the core area of the other—they would engage in hostile displays directed at one another. These encounters happened less and less often during the ensuing couple of years as the males of each subgroup avoided contact with the other.

By 1973, two years after Goodall had observed the first incipient fission of her community, she recognized two distinct communities: Kasekela in the north and Kahama in the south. They occupied separate but slightly overlapping territories, and each community's males began to treat excursions to the overlap zone as boundary patrols. By the following year, violent encounters began to take place between the now-rival communities. The territory of the northern Kasekelas expanded while that of the smaller Kahamas contracted. By the end of 1974, the Kahama chimps were squeezed into a small patch of forest of less than four square kilometers (there was another, unstudied community called Kalande farther to the south hemming the Kahamas in). Several months later, the four Kahama males restricted their movements even further, rarely leaving a tiny patch of their former territory estimated at less than two square kilometers. They must have believed they were in dire straits, and they were. From 1974 to 1977, the Kasekela males made repeated incursions into the Kahamas' territory, sometimes brutally attacking Kahama males. The Kahama males were attacked and killed one by one until all were dead. The Kasekelas then subsumed the former Kahama range into their own.

The attacks were shockingly violent, all the more so because the killers and victims had only a few years earlier been allies. Lacking claws or carnivore-like canine teeth, the marauding males nonetheless inflicted severe bite wounds, and they engaged in protracted beating with their fists that resulted in apparent broken bones and internal injuries. The attacks were carried out by two to six Kasekela adult males, with adolescent males sometimes involved too. Ironically, not long after the four-year "war" was over, the large Kalande community farther south began pushing its range northward, and some of the range that had recently been won by the Kasekela chimpanzees was lost. In recent years such battles have also been witnessed to the north of the Kasekela community, where the small Mitumba community is wedged between the Kasekelas and the village beyond the northern park border.

Although the Kasekela males seemed to seek out and attack rival Kahama males, females were targeted too. Intercommunity conflicts in chimpanzees can involve attacks on older females, often with their offspring in tow, and more rarely on younger childless females. Younger adult females without infants may face aggression from the adult females in the new community in which they settle. The males, on the other hand, welcome them with open arms. But older females who find themselves in the path of patrolling rival males are often the targets of vicious attacks. These were well known long before anyone suspected that intercommunity battles between males existed. These attacks may target a female, but her infant is a more frequent victim. Unable to hurt the mother severely enough to prevent an escape, the males settle for taking her infant.[35]

The account of community fission followed by systematic annihilation at Gombe is by no means an isolated incident. Down the lakeshore, Nishida had noted early in his research that male chimpanzees of adjacent communities had hostile relations. Mahale M group and K group had overlapping home ranges, but during the dry seasons of the 1960s and 1970s, M group moved north in large parties into the overlap zone.

K group usually responded by moving away into the far north of its own range, avoiding the more powerful M group males. Occasionally K group did not migrate seasonally away from M group, and then territorial encounters ensued. During some of these encounters, M group males attacked both males and females of K group. In the 1970s, paralleling the "warfare" happening one hundred kilometers away at Gombe, most of the K group males disappeared; by the early 1980s, they were nearly gone. All had last been seen in good health. Although the killings were not observed, Nishida and the other Mahale researchers believed that most if not all of the K group males were killed by M group.[36]

In Kibale National Park, both Kanyawara and Ngogo communities have experienced lethally violent attacks on males from neighboring communities. These were usually but not always attacks on adult males. Juveniles have occasionally been targeted, and females in the company of males who were victims have also been attacked. Far to the west, in Taï National Park, multiple overlapping communities have been studied for decades, and a slightly different picture of intercommunity aggression has emerged. First, while attacks occur, it took more than twenty years before a lethal attack was observed. Boesch has described a variety of attack strategies, in which the males of the study community appear to adopt particular directions from which to charge at the rival males at a territorial border. Sometimes they launch a frontal attack; at other times a flanking attack or a rearguard action is called for.[37]

If these sound like troop movements on a primitive battlefield, that is exactly the way Boesch sees them. He even claims that the rearguard action is an act of intentional deception, in which the attacking males leave some members to the rear to call loudly and give an appearance that the attackers are not as close to contact as the forefront really are. One can question whether, in the dense forest in which these intercommunity encounters occur, an observer can see such detail in the chaos. The most common form of attack at Taï is the one that we consider the norm for all chimpanzee communities: the "commando" raid, in which a few adult males launch an incursion deep into enemy territory and attack any lone males or females they find.

Following the systematic killing of K group males at Mahale, the remaining females of K group actively sought to join the M group community. At both Gombe and Mahale, males sometimes attack and severely injure females from neighboring communities. This seems counterintuitive if one of the reasons for male incursions is enhancing mating success. We might expect males to either coerce or embrace new females whether they are young and childless or older. It could be that it's more desirable to simply expand one's territory without having to add more mouths to feed, regardless of sex. Research at Gombe by Jennifer Williams and her coauthors showed that females with larger home ranges have shorter intervals between births, suggesting that larger ranges provide more food and are therefore a valuable asset worth excluding others from.[38] Gauri Pradhan of the University of Zurich and colleagues argue that killing females rather than allowing or coercing them to enter a new community is reproductively advantageous under certain circumstances. High-ranking males have a greater potential payoff from attacking females (and their infants) to exclude them from new territory gained because they already have a wide choice of females in their own community. Low-ranking males, on the other hand, should be open to inviting any and all new females in.[39]

Explanations for Lethal Territoriality

Recent studies of intercommunity aggression have taken a more experimental approach in order to understand the dynamics of chimpanzee intergroup conflict. In a series of papers in the 1990s, Wrangham and colleagues developed an idea first suggested by Goodall: we should expect intercommunity attacks to follow patterns of an imbalance of power. Attacks should be expected when the disparity in numerical strength favors the attacking group to the degree that the risk of injury to the aggressors is least. When ten chimpanzees happen upon a lone member of a rival community, they know they can attack with a low risk of serious injury to themselves. The same pattern may describe hunter-gatherers or soldiers on patrol in a war zone; numerical superiority has always been a key variable in the equation of when and whether to fight. Lethal chimpanzee raids do not follow from a smaller or larger conflict. They are usually surprise attacks aimed at leaving victims seriously wounded or dead.

Wrangham viewed intercommunity attacks as the product of an imbalance of power, with males motivated by the quest for dominance of their community over a neighboring one. He has often taken a resource-oriented view of chimpanzee society, in which male foraging parties defend valued fruit trees. There is nothing more dangerous to a lone male chimpanzee than foraging in an overlap zone patrolled by males from an adjacent community. Of course, in a fission-fusion society, chimps will sometimes be alone. The larger the party, the better able it is to control food that is also prized by neighboring communities. Forests with a richer abundance of fruit should, according to Wrangham, feature large chimpanzee parties and more aggressive intercommunity encounters. This is because large disparities in party size leads to more frequent encounters as one side or the other perceives an imbalance of power.[40]

Wilson and his collaborators working at Kanyawara examined the factors influencing male strategies of intercommunity attacks. They found that encounters were most likely when males were far from the core area of their range and when they were eating fruits that were most abundant in that area. A few key fruit species that were sought after by members of adjacent communities led males into these high-risk areas. When the Kanyawara males heard calls from their neighbors, they approached rather than retreated when they were numerically stronger than the neighbors, based on the number of distant calls heard by the human observer. Encounters were influenced by the presence of desired foods, but actual attacks were predicated on the force of strength of males present.[41]

Gathering detailed information on intercommunity attacks takes decades, because the attacks are infrequent and often not clearly observed. So Wilson and his colleagues used loudspeakers to broadcast an unknown lone male chimpanzee's calls to the males of the Kanyawara community, tricking them into believing an intruder was near or inside their territorial border. He then measured the response. One might think that chimpanzees, as intelligent as they are, would not be easily deceived by loudspeaker broadcasts of calls. They might be expected to realize quickly that they're being pranked, in the same way that you might be fooled the first time you heard a recorded voice of a stranger in your attic but would know better by the second or third time. Wilson argued that chimpanzee playbacks are no different from those used successfully in experimental field studies of birds, lions, and other primates.

The response to playback calls depended on the number of males hearing the call. When three or more males heard one intruder's call, they responded aggressively, approaching and calling back. But parties with fewer males took a more circumspect approach. They refrained from calling and didn't necessarily approach the speaker. In other words, the males estimated the strength of the perceived enemy chimps based on the number of calls they heard, presumably to gauge the risk of confronting the intruder. The location of the speaker did not influence the behavior of the chimpanzees that heard the faked calls. Although the sample size was small, it lends support to the idea that males are eager to fight to defend females, territory, or both.

Why Do Males Chimpanzees Attack Their Neighbors?

The imbalance of power between parties of two adjacent chimpanzee communities provides a strong explanation for the immediate causes of intercommunity attacks. But a deeper evolutionary explanation is not so widely agreed on. Watts, Muller, and their colleagues argue that intercommunity territoriality at Kanyawara and Ngogo is all about enhancing and defending food resources for the sake of increased fertility, well-nourished females, and higher overall survivorship of offspring. If attacks were for the sake of reducing coalitions of rivals, then females ought to be attacked regularly too. Food defense explains territorial incursions if neighboring females change or shrink their ranging patterns to avoid further attacks, thereby allowing the attackers greater monopoly over the food trees sought in the territorial overlap zone.[42]

Williams and her collaborators showed that food resource defense was also the most important factor in attacks at Gombe. They cited male attacks on both females and males from adjacent communities as evidence that incursions are more about controlling the food supply than about capturing or controlling females. They found that territorial expansion did not bring with it an increase in the number of females in the community, but it did lead to higher reproductive success for the resident females. Based on circumstantial evidence—more time in larger foraging parties and more encounters with local females—the researchers believe that larger territories may increase the availability of preferred foods. Williams and her colleagues considered male territoriality to be all about maintaining safe feeding grounds for their females and offspring while also protecting them from contact with rival males.[43]

Most studies, including Goodall's own work, have either shown or assumed that the number of adult males in a community is directly related to the size of the territory. More patrolling males should capture more land and also intimidate weaker rival communities to withdraw from their own lands to avoid disastrous encounters. But Williams and her coauthors failed to find any association between the size of parties at Gombe and their ability to control territory. The same lack of association is seen at Taï. Mitani and his colleagues at Ngogo found that intergroup aggression did lead to territorial expansions, as at Gombe. The Ngogo community is so large and powerful that the males may simply be unchallengeable and thereby force their neighbors to withdraw.

It seems clear that the evolutionary force driving chimpanzee lethal territorial attacks is sexual selection. Behaving violently toward neighbors is adaptive because it (1) eliminates rival coalitions of males along with their territories, (2) obtains females from those eliminated communities, and (3) enhances access to desired foods. Teasing apart these three factors will take much more field research, since intercommunity attacks don't occur very often. After a half century of observation, we can say with certainty today that lethal aggression is a strategic, adaptive, and routine aspect of chimpanzee social behavior.

Killing Infants

In the animal kingdom, the killing of infants was long seen as pathological. But in a wide range of mammals, males attack and attempt to kill the infants fathered by rivals. Among lions, squirrels, and mountain gorillas, to name a few, infant mortality due to male infanticide is not only a major cause of death but also a male reproductive strategy. In the 1970s, primatologist Sarah Hrdy set off a firestorm of academic debate about infanticide when she reported male hanuman langur monkeys in Mount Abu, India, invading new groups, ousting the resident male, and attempting to injure or kill infants in the group. Hrdy followed the new wave of Darwinian thinking about animal social behavior—which is today the orthodoxy—and interpreted such killings as males seeking to eliminate the genes of their rivals. Additionally, females that lost their infants began cycling again, allowing the new resident male to get started on his reproductive agenda without waiting for the resident male's offspring to mature. Some primatologists claimed Hrdy's langurs were behaving pathologically due to the stress of living in a crowded urban environment. Then more researchers witnessed langur infanticide in settings more pristine than Mount Abu. It took years of observation and multiple further field studies, but eventually Hrdy's ideas were confirmed.[44]

Over the ensuing decades, voluminous data on infanticide in a range of primate species have been collected. The past three decades of research and theorizing about primate infanticide have brought primatologists to an understanding of some general rules and evolutionary outcomes of the behavior. Infanticide within a community is a relatively rare occurrence among chimpanzees, and recorded cases are perplexing. Wilson and colleagues compiled forty-five instances of intracommunity infant killing in nine communities in what amounts to more than two hundred total observation years studying wild chimpanzees.[45] They found at least forty additional infants to have been killed by enemy males during intercommunity attacks across fourteen different sites, perhaps because they are more easily captured and injured than their mothers. It might make reproductive competitive sense for adult males to kill male infants that could eventually immigrate into their community, perhaps accompanying their mothers. On the other hand, such a male immigrant would grow up to help the community's males during patrols and hunting. Wrangham and his collaborators pointed out that intercommunity attacks against infants appear to target males, but the sample size is too small to make much of that finding. We don't know the cost, from a male's perspective, of having another male mouth competing for food versus the benefit of having that new male group-mate to help him someday acquire access to food and females.

Cases of infanticide within a community are rare enough that no clear pattern has emerged to explain them. Carson Murray and her colleagues analyzed thirteen intracommunity killings in search of a pattern. Infants from as young as three weeks up to two years old were victims. Most were the offspring of high-ranking mothers. Of the twelve victims whose sex was identified, ten were males. Likewise, most of the attackers were males.[46]

Females sometime commit or attempt to commit infanticide. At Gombe, Fifi, the high-ranking female who was as doting and nurturing a mother as one could imagine, was also known to try to steal infants from their mothers with the appearance of harmful intent. On one occasion at Gombe in the 1990s, Fifi attempted to grab Gremlin's newborn baby repeatedly, and when she failed to wrest the infant from Gremlin's nervous grip, Fifi enlisted reinforcements. Research assistants watched as she tugged another female, Patti, by the arm, attempting to pull her into the fracas until Gremlin wisely departed the area, her infant cradled tightly in her arms.[47] The most infamous baby killer of all was Passion, who went on a veritable infant-killing spree at Gombe in the mid-1970s when she and her adolescent daughter Pom began to steal and eat the infants of fellow females. They killed and ate at least three, and possibly as many as ten, infants over a four-year period. Pom's cooperation played a key role, enabling Passion to capture infants she would likely not have been able to kidnap alone.[48]

Whether we're considering males or females as infanticidal killers, we have to distinguish between infanticide and cannibalism, since they may have utterly different motives. Passion's murderous ways ended when she died, and no female remotely like her has been observed at any study site. Whether she was behaving in a way that reduced her female rivals' reproductive output or perhaps had a nutritional deficiency that led her to kill baby chimpanzees—maybe they were easier for her to prey on than monkeys would have been—is unknown. Most infant killings, whether by adult males or females, involve the consumption of at least part of the tiny carcass afterward. The motivation behind the act might be nutritional. It could alternatively be that the goal is to eliminate a future food competitor for the same resources. Infanticide could be reproductively adaptive for adult males, who might be eliminating a rival's genes. That would require them to be able to distinguish their own offspring from nonrelated infants. It's not clear whether infanticidal chimps can do this. This has made intracommunity infant killings by males at Gombe, where it has happened at least once, and Mahale, where it has occurred at least seven times, difficult to explain.

Making Up

We—the public, the media, and scientists themselves—are riveted by violence. We are far less riveted by an equally important component of aggression: the reconciliation that occurs afterward. Without the smoothing over of disputes, the web of relationships that are the fabric of primate society would dissolve. That includes human societies, where rituals large and small (plus a court system in some societies) function to restore harmony after verbal or physical aggression has occurred. Peacemaking is every bit as important as aggression to any social animal.

In an earlier era, we assumed that when social animals have disputes, it causes them to avoid one another afterward. But that is not the case. The first study of postconflict resolution in chimpanzees, by Frans de Waal and Marc van Roosmalen, showed convincingly that chimpanzees were actually *more* interactive in the minutes after a conflict.[49] Why they would bother to seek reconciliation at all may be crucial information for our understanding the costs and benefits of being social, which in turn could answer questions about our own origins as an intensely social species. But there are layers to reconciliation, since the benefits to both aggressor and receiver may be quite different depending on dominance rank and other aspects of the social context.

There have been many studies of peacemaking among nonhuman primates over the past two decades, including those of chimpanzees. These have almost all been conducted in zoos or laboratories, where behaviors are most easily observed. When a fight breaks out, the sequence of events may unfold within seconds, and, even in a captive enclosure with multiple observers armed with video cameras, details are easily missed. In a dense, dark forest, very little may be seen of either the conflict or the reconciliation after it. De Waal is a pioneer of studies of reconciliation. His relational model of the costs and benefits of conflict and conflict resolution was the basis for much of the research on the subject since. Individuals have three mutually exclusive options when competing with a rival over food, mates, or other valued resources. They can fight, they can ignore the rival, or they can actively avoid aggressive interaction. As you might expect, as the value of the object of competition increases, it tips the scale of decision making in favor of risking aggression. At the same time, the risk of permanently fracturing a valued relationship may increase when the aggressive conflict intensifies. So, de Waal reasoned, the odds that a primate will escalate an aggressive encounter depend on the odds that the relationship can be repaired later through reconciliation.[50]

Numerous captive studies of primate reconciliation have tested de Waal's ideas. Comparable studies of postconflict behavior in the wild are badly needed to see the behaviors as they occur in the context in which they evolved. There have been only a handful of studies of postconflict behavior among wild chimpanzees, and some of these have only observed a small number of reconciliations. Kate Arnold and Andrew Whiten observed 120 postconflict behaviors in more than a year of observation of the Sonso community at Budongo. They found that reconciliation, when it did occur, took place almost immediately after a dispute.[51] My doctoral student Jess Hartel conducted perhaps the most detailed study to date of chimpanzee reconciliation in the wild, among the Kanyawara chimpanzees of Kibale National Park in Uganda. She found that the chimpanzees had a plan; they evaluated the costs and benefits of reconciling after a dispute versus not. Preferred social partners—friends—played key roles in determining whether repairing a relationship immediately was worthwhile. One-sided relationships ("I need you more than you need me") were the best predictors of when reconciliation would happen. But relationships between mutual partners ("We're both close friends and we know it") were better predictors of the form that reconciliation took and how it varied in different social contexts. And, as you would expect from the chimpanzee male-bonded society, males were more active reconcilers than females.[52]

Together with the detail provided by captive studies, field studies give us insights into the ways in which chimpanzees maintain social harmony. Roman Wittig and Boesch tested de Waal's relational model among the Taï chimpanzees. The Taï chimps demonstrated the same variety in judgment about social situations that we find in people. Some appeared to misjudge the situation, being too aggressive when the reward really didn't warrant the risk of fracturing the relationship. Others were all too cautious in their approach to such encounters. But overall, the Taï chimpanzees followed the rules of conduct predicted by de Waal.[53]

Wittig and Boesch also showed that among the Taï chimpanzees, the choice of postconflict behaviors has a lasting effect on the nature of one's relationship with a social partner. After an aggressive dispute, a chimp attempted to reconcile mainly when doing so would not risk further antagonizing his or her social partner. As you would expect, it was harder and took longer to reconcile after intense aggression than after minor spats. And when food resources were abundant, reconciliation after food-related fights was less likely, presumably because that abundance was easily shared, leaving little incentive for either party to reconcile.

As we've seen, chimpanzees turn nuanced political behavior into an art form. Interactions are often far more complex than merely actor versus recipient. Bystanders and "friends of friends" are often involved during or after the fact. Wittig and Boesch examined the role of bystanders in the postconflict phase of aggressive encounters among the Taï chimpanzees. As we see in ourselves, chimpanzees "make up" with the rivals of their friends at times when neither the friend nor the aggressive rival is in the right mood to do so. Testing various hypotheses to explain the role of bystanders, the researchers concluded that friends play a key role in repairing the relationship status of the aggressor and the friend that existed before the unpleasantness. It also seemed to grease the wheels for reconciliation between the two antagonists in the subsequent bout of nastiness, even if the friend wasn't present the next time. Most importantly, the authors observe, the process of reconciliation provides a window into the chimpanzee mind. The individuals involved have to have a long-term relationship, and the friend must be cognizant of the nature of that relationship. They're aware of the intricate web of alliances and rivalries among community members, and their social intelligence can provide tangible benefits in the form of more readily repairing rifts that happen daily. They are, in other words, very much like us in navigating the web of friendships, rivalries, and relationships of need and convenience in the course of their daily lives.

CHAPTER 5

Sex and Reproduction

> 1963. . . . For the next week, Flo is followed everywhere by her male retinue. If she sits or lies down, several pairs of eyes swivel in her direction; when she gets up, the males are on their feet in no time. Whenever there is any kind of excitement, the adult males one after the other copulate with Flo. There is no fighting; each takes his turn.
>
> Jane Goodall, *The Chimpanzees of Gombe* (1986)

A FEMALE CHIMPANZEE experiences her first sexual swelling at around eleven years of age. Adult males suddenly see her in a different light and she begins to mate with them. A year or two later, she will likely migrate from the community in which she was born. She may spend months in a neighboring community before transferring back home. Eventually, by about age thirteen, she will have settled into another community in which she will probably spend the rest of her life. By her midteen years, the female will have conceived an infant in her new community and given birth. Her first baby has a less than 50 percent chance of surviving to adulthood. She will nurse the infant for four years and provide psychological support as only a mother can for several years after that. The female will give birth every five years until she dies, and although her fertility will decline in old age, she will not experience the abrupt menopause that human females undergo. These basic parameters of a female's life cycle are well established. But new revelations, combined with existing information about sex and reproduction, have created a new picture of chimpanzee sexuality and its implications.

Swellings

A female's reproductive career may span several decades, but there are hurdles at each life stage that threaten reproductive success. The earliest primatologists saw female primates mostly as objects of male desire and vessels for carrying babies. Since the 1970s we have recognized that chimpanzee females are active strategists who plot their own reproductive agendas. Beginning at sexual maturity, females with sexual swellings are very sexually active. David Watts observed an average mating rate of three and a half copulations per fertile female per hour at Ngogo. One female mated sixty-five times, with eighteen different males, in an eleven-hour day.[1] That's one sexual encounter every ten minutes.

Sexual swellings have evolved several times independently in the Primate order. They are boldly visible in some Old World monkeys, chimpanzees, and bonobos; much less so in other primates.[2] The anogenital swelling encompasses the anal and vulval areas. It's a massive appendage that swells with up to a liter and a half of fluid for twelve or thirteen days of a sexually mature female's thirty-five-day menstrual cycle (the cycle shows wide individual variation among females, from two to eight weeks). Ovulation, if it occurs, is the fertile time window roughly in the middle of the swelling cycle; the swelling is most distended at this time. Males are intensely curious about and excited by both the sight and odor of female swellings. Mariska Kret of Leiden University and Masaki Tomonaga of Kyoto University argue that a male chimpanzee's ability to recognize the sexual state of a female's swelling is as important to his social life as facial recognition is to ours.[3] Because of the difficulty of doing systematic research on hormones in an African forest, captive studies taught us most of what we know about a female's sexual swelling until the past decade or so. The swelling itself is strongly influenced by the action of estrogen in the days just before ovulation, while the loss of the swelling a few days later is mediated by a rise in progesterone.

Swellings inflate over a period of days. In the wild, female chimpanzees normally mate only when they have full or nearly full swellings. Captive females will sometimes mate when barely swollen, perhaps from enforced proximity to males, or just boredom. It has also been reported that males may prefer mating only with very swollen females. Janette Wallis of the University of Oklahoma College of Medicine showed that males inspected swellings less frequently but mated more often as ovulation approached. Males behave as though they are aware that ovulation occurs during only a fairly narrow time window in the ten-day swelling. Females swell not only when they're ovulating but also during pregnancy and when nursing an infant. In a separate study, Wallis showed that, at least in captivity, pregnant females mate at a higher rate than cycling females, particularly early in the pregnancy. Males are excited visually by the swellings, even if there are no clear olfactory cues that the female may be fertile at the moment. Cycling females may be more selective about whom they mate with than pregnant females, who stand no risk of impregnation by a less than desirable mate.[4] Sarah Hrdy and others have suggested that female sexual behavior during pregnancy is adaptive in that the female may confuse paternity of her future offspring in hopes of mitigating aggression from males toward the infant.[5] Females also reap benefits while swollen, such as a rise in social dominance and therefore better access to the company of high-ranking social partners and better access to food, including meat.

Mating peaks a week before the swelling begins to detumesce, and this peak coincides with peak estrogen levels. Anyone who has watched chimpanzees in the wild or in captivity has noted that males are visibly more

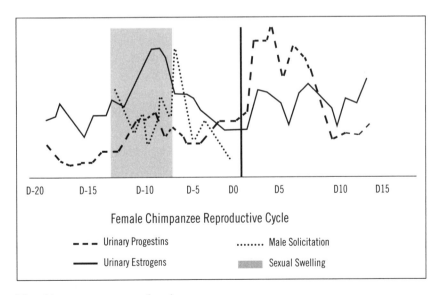

The chimpanzee menstrual cycle.

interested in swollen females that are ovulating. Nonovulatory swellings during pregnancy or lactation elicit some interest but rarely an extreme state of sexual excitation. This is not to say that those females don't mate; just that they do not attract a large party of eager males, nor an alpha who must spend his energy and time fending off other suitors. It's clear that whether the cue is visual, olfactory, or a combination of the two, males seem aware of a female's cycle to some extent. It's also clear that some females of prime breeding age bearing large swellings don't attract very much attention from males. This may be due to a variety of factors that influence whom males choose to mate with. This is an issue that has been largely overlooked in primate research for the past thirty years in favor of a nearly exclusive focus on female mate choice.

On the other hand, females who are highly desired and swollen attract so much attention from the males that pandemonium may ensue. I have followed parties of males who sat beneath trees, all their penises erect, while the female and the alpha male sat high above. Males would race up the tree toward her if the alpha turned away for a minute. A fight would ensue, and the alpha would resume his days-long task of keeping all the other suitors at bay.

Females appear to use their swellings to indicate their desirability as a mate. The size of the swelling itself is the likely signal and is influenced by the state of the ovaries. Melissa Emery Thompson of the University of New Mexico, Albuquerque, and Patricia Whitten of Emory University reported that among captive female chimpanzees, swelling size was greatest during the ovarian follicle-release and ovulation stages of the cycle. Males cluster around her in hopes of mating during the time window just before ovulation.[6] A study by Tobias Deschner of the Max Planck Institute for Evolutionary Anthropology and his collaborators established that swellings increase in size leading up to the approximate time of ovulation. There is typically a period of several years between the first swelling a female experiences and her first pregnancy—a prolonged period during which the adolescent female is unlikely to conceive. During this time the swelling is not necessarily a reliable indicator to males of which females are fertile.[7] Emery Thompson and Whitten noted that swelling size alone is still a generally reliable signal of both a female's ovulation and the likelihood of her conceiving.

The swelling does not necessarily indicate the overall genetic quality of a female. That it visually signals the immediate state of fertility is rare among mammals. In most species, males advertise their quality with ornamental anatomy or aggressive behavior. The hormonal fluxes that a female undergoes each cycle will change over the course of her adult life span, even for species such as chimpanzees in which there is no marked menopause. So the swelling is a billboard advertising her reproductive quality at a given point in time, both within her cycle and also within her prime reproductive years. The information that it conveys to males is important but likely imprecise enough that they cannot pinpoint the exact timing of ovulation, which leads to the frenzy around females during their swollen periods.

The swelling is a large, delicate appendage that is prone to infection, parasites, and injury during male scrums over a female. A female with a swelling attracts males, which allows her to watch males compete but also risks injury to herself. A swollen female may lose precious time needed to search for food because of the attention her swelling attracts. All of this suggests that the swelling must confer a well-deserved reproductive benefit to a female for all the trouble it causes her. The interplay between female mate choice and male mate choice is an area of research that we need to explore far more deeply in coming years.

The dimensions of the swelling also have to be right for the insertion of a penis. At maximum swelling, the depth of a female's vagina increases dramatically from its usual state. A male chimpanzee's long and very thin, saber-like penis is adapted to deep penetration of a swollen vaginal area. Captive studies by Alan Dixson and his colleagues have shown that size matters; males vary in their ability to successfully achieve deep penetration. Chimpanzees and the males of many primate species possess a baculum—a penis bone—which humans lack. Dixson argues that the baculum is the product of sexual selection and that its length relative to body mass is associated with how many minutes copulation lasts. It may also serve to protect the male's urethra during mating and to stimulate the female and thereby enhance the odds of fertilization. Male chimpanzees, like a few other primates, produce a copulatory plug—a hard, gelatinous mass that is deposited against a female's cervix. This may prevent loss of sperm that have been deposited or it may prevent the sperm of other males that mate with that female afterward from reaching the cervix.[8]

Anyone who's seen a male chimpanzee has noted one—really two—of his most prominent features. They have one of the highest ratios of testis size to overall body size of any primate. According to Dixson, the combined weight of the testes may be 120 grams—more than a quarter of a pound—in stark contrast to the less than 30 grams of testes in a male gorilla with three times the body weight. The long-standing assumption has been that in chimpanzee society, in which a single female may mate hundreds of times with many males while she is ovulating, maintaining exclusive access to that female may be a futile effort even for the alpha. Therefore, competition may occur at the level of the sperm in addition to that of the males themselves.[9] Alexander Harcourt of the University of California, Davis, and colleagues showed decades ago that large testes-size-to-body-size ratio in primates is associated with a multimale polygynous or promiscuous mating system. No one has ever conclusively shown, however, that sperm competition in chimpanzees actually occurs.[10]

Chimpanzees are an odd species reproductively, because females reproduce very infrequently but are highly promiscuous when fertile. Males have very few chances to father a baby with a given female during her lifetime (several times fewer than a human male with his female partner), even though the female's periovulatory status is visible to males. But Emery Thompson points out that males are nevertheless challenged to know exactly when to maximize their efforts to mate, because swelling cycles vary a great deal from female to female and from one cycle to the next. Males may use subtle changes in female swelling size as a visual cue, and they likely also use olfactory cues that are still poorly understood. But they are still probably unsure exactly when a female experiences maximum fertility. This is what creates all the male competition, and thus opportunities for female mate choice.[11]

One additional way in which a female can signal her attractiveness is with copulation calls. Females call—basically scream—when copulating, perhaps to advertise to the males at large that they are sexually available and available to be competed over. The call is under some voluntary control; on patrols and in stressful situations, females may visibly suppress the call so as not to attract undue attention to themselves or the party. Simon Townsend and his colleagues at the University of St. Andrews have investigated the role that copulation calls play in chimpanzee society. They found that females gave copulation calls even when they were not ovulating. This suggests that a female uses the call to manipulate males around her rather than honestly signal to them about her reproductive status. When females mated with high-ranking males, they tended not to call when high-ranking females were also in the vicinity. Again, the copulation call is a signal, and it is best not to attract attention to oneself in certain mating situations. Townsend and his coauthors examined the contexts in which the call was given among Budongo chimps, and also its acoustical structure. They predicted that for the call to be highly effective at attracting and inciting competition among males, the males should be able to identify a female by her copulation call. They found that copulation calls inform males about the identity of the female caller, but they don't reflect her ovulatory or nonovulatory state.[12]

Meanwhile, Brittany Fallon, also of the University of St. Andrews, and her colleagues found that Budongo females adjusted their copulation calls depending on the context of females and males around them, and also with regard to their reproductive state. As the number of cycling females present increased, females increased their copulation call rate to compensate. This is presumably a reflection of female competition. Young females who had yet to produce a baby gave more copulation calls than older females who had already given birth; these newbies perhaps more badly needed to signal their availability. As in Townsend's study, the rank of males in proximity was a strong influence on copulation calls, as females were more likely to call during mating within earshot of high-ranking males.[13] These results don't mean that female chimpanzees consciously make subtle adjustments in their vocal behavior to suit the social situation. Differing levels of male interest may be a proxy cue to induce female calling behavior. But given the cognitive sophistication of the chimpanzee, it shouldn't surprise anyone that they have some volitional control over their vocal behavior, even during mating.

Choosiness and Choice

The excitement that a female chimpanzee's swelling causes among males has profound implications. Advertising her reproductive status allows her to attract a gathering of males who vie for her for several days. She can pick and choose, to some extent, the male she mates with and perhaps influence the paternity of her offspring. She can be *simultaneously* promiscuous and selective in a way that few other primate species are. But there are limits to her choosiness. We are still unraveling the complexities of chimpanzee mating and will be doing so for decades to come, but some aspects have become well understood in recent years.

A female who wants to copulate with a particular male will be swollen and already attracting a great deal of attention from him. A male may be possessive of her, or she may be in the company of many males when she positions herself in front of one and turns to offer her swelling. Rarely does she take this step without some indication of interest from a male. It could be a male's branch-shaking or ground-thumping-with-knuckles gesture, which symbolically signals his desire to mate. It could be his gaze directed at her or simply that he is sitting near her with an erect penis. These invitations are ignored by females at their peril; refusing him may provoke aggression from the male. But this does not mean that the female always solicits males who perform those invitations. And sometimes a female will present her swelling to the male without any visible signal from him at all.

Female choosiness and choice are not the same thing, biologically speaking. To be choosy means simply being discriminating, and it usually refers to the tactics a female uses when deciding whether to mate and with whom. Primatologists use "choice" to refer to the deeper, evolved reasons that females choose a male. These can include physical attributes that are signals of genetic quality—body size, canine tooth size, healthy appearance, and so on—or behavioral qualities. Prime-age males may be the best-equipped fighters physically, but in chimpanzee society both social intelligence and one's bloodline matter enormously. Teenage males can't often challenge high-status males, but females nonetheless mate furtively with them, behind bushes or out of sight of the alpha. Females also may prefer the company of older, past-prime males, who offer the same genetic traits that helped propel them to high rank years earlier but are now less aggressive or coercive to them. Nonetheless, studies of paternity among wild chimpanzees have generally found that males at or near the top of the dominance hierarchy father most of the babies. So either a female's ability to select the male of her choice is limited by male dominance or females overwhelmingly prefer to mate with dominant males.

As Goodall observed in the vignette that opened the chapter, males don't always fight to monopolize females. In some cases they allow matings by other males to happen right under their noses. In Mahale Mountains National Park, Akiko Matsumoto-Oda addressed this seeming paradox. She found that females used two strategies. They mated more frequently when maximally swollen but did not expand the pool of males they mated with. As the estimated day of ovulation approached, females restricted their mating to the high-ranking males, with whom they also associated and groomed most often. The twin strategies of mating very promiscuously (a few matings with each of many males) and less promiscuously (many matings with each of a few males) were used sequentially in the same swelling cycle.[14]

Why would a female mate with so many males when the odds of conception peak? Matsumoto-Oda argues that it's due to the very low odds that any one mating results in fertilization. She estimated nearly six hundred matings per conception for Mahale females. If a female limited herself to one or two males, their diminishing returns of ejaculate and sperm production in the course of several hours might further limit her ability to ensure herself a conception. Mating with many males would ensure that at least some of them are mating with good odds of fathering her offspring.

At Ngogo, Watts reported similar findings, but the enormous size of the community there and the large number of males played a role in dictating female sexual behavior. As at other sites, females mated most often around the time of ovulation; previous mothers mated more often than females who had never given birth. At Ngogo, fully swollen females mated with more males than at other sites, presumably because the available pool of males was much larger.[15]

How effective are females in choosing the mate they prefer? Rebecca Stumpf and a group of collaborators from the Max Planck Institute for Evolutionary Anthropology examined female mate choice among Taï chimpanzees. They found that females had substantial freedom of choice despite the constraints of the male dominance hierarchy. Females and males both took the initiative in mating, and in each case the opposite sex sometimes cooperated but at other times rejected the potential partner. When females reject male suitors, they may face his wrath and end up mating with him to avoid aggression.

Stumpf and colleagues found that females initiated one-quarter of all sexual interactions, and males accepted those advances nearly 80 percent of the time. Higher-ranked females were more likely to mate with the males they chose than low-ranking females were. In general, higher-ranked males were more likely to look favorably on a female's advances as well. Males took the first step three-quarters of the time, and females agreed to mate with them about 70 percent of the time. When females attempted to avoid mating with a particular male, they were successfully able to evade him nearly 70 percent of the time. A swollen female near her estimated time of ovulation was better able to avoid a male's advances if that male was not the alpha but the alpha was sitting nearby.[16]

One might ask why males, given the tiny amount of time invested in mating and the trivial cost of their sperm, would ever turn down a female. Social politics among the males may be the explanation: subordinate males may be unwilling to mate in front of an alpha for fear of reprisal. And one might also ask why females ever turn down males, since the odds of conception in a single mating are so low. In fact, females at Taï reject some males at the point of their cycles in which ovulation is likeliest, but they take the initiative with those same males when the odds of conception are nearly zero. In other words, they are most selective when discrimination matters most. They are also cleverly playing reproductive politics, as they strengthen bonds with males and perhaps discourage aggression from them toward their future offspring. Stumpf also found that while every Taï female took the mating initiative with males, there was much variation among females in the degree. Overall, females behave strategically during their fertile window, and more research about that window of her swelling cycle is likely to tell us much more about the nature of female choice.

Other studies have found varying results on the topic of female choice. At Gombe, Wallis found that females were only about half as successful as Taï females at warding off male advances.[17] And at Mahale, Toshisada Nishida reported that male sexual advances were successful far less often than at either Gombe or Taï.[18] It may be that sites with a larger number of powerful males have different mating dynamics because the odds of a male being challenged, or a female being assaulted for refusing, are greater. And adolescent males may adopt altogether different mating tactics from adults. At Ngogo, Watts found that adolescent males mated more with females who had never given birth and who mated mainly outside their fully swollen periods. In other words, they either opted for a suboptimal mating pattern compared to their elders or were forced into those less desirable matings.[19]

On a community-wide scale, female reproductive cycles play an enormous role in the lives of all members. Because chimpanzee communities typically contain ten or more adult females, and births occur in every month of the year, there is no apparent synchrony of swelling cycles.

Whether this asynchrony is the product of natural selection is an open question. There are various reasons that female primates may want to schedule their fertility. For some small species that are eaten by many predators, giving birth at the same time as other females may lessen the chances that one's offspring is the victim of predation. Such birth clustering effectively swamps the predator, and the pattern occurs in many small mammal species. There are two possible factors at work here: the timing of ovulation and the timing of births many months later. Being fertile in a given season may be selected because there is an optimal season to give birth due to food supply, or an optimal season in which the baby is weaned and can find its own food. Such is the case for some primates living in very seasonal climates with cold, harsh winters.

On the other hand, it may be advantageous for a mother to ovulate at a time when no other female is. This would decrease female competition for the best male—though highly promiscuous mating in chimpanzees makes this less of a factor. It may also lower the risk of being coerced to mate with males who are not the desired mating partner, because the alpha would have no other females on whom to focus his time and energy. In Mahale, Matsumoto-Oda and her colleagues found that female birth staggering occurs. Another effect, likely a by-product, of cycle asynchrony is a reduction in the overall birth rate. Mahale females had lower birth rates in years in which births were most staggered. When a female was the only cycling member of the community, she mated more frequently than when males' attentions were divided among multiple swollen females. It was not surprising that the alpha could best monopolize matings with a swollen female when no other females were swollen.[20] Kimberly Duffy of the University of California at Los Angeles and her collaborators found that male chimps at Kanyawara engage in social bartering over access to females. Low-ranking males' mating success was closely tied to their support of the alpha. He tolerated their attempts to mate as long as they paid him obeisance in the form of grooming and coalitional support when the alpha needed it.[21]

Group versus Consort

Male chimpanzees employ two general mating strategies: opportunistic group mating and consortships. Group mating, as Goodall first described it, is what I've described thus far. During the fertile portion of her cycle, a female associates with a number of males, all of whom are interested in her. Although aggression does break out, often in the context of redirected sexual frustration, there is also a great deal of tolerance among the males sharing the opportunity to impregnate the female. The difference between an ovulating female who incites chaos and one who does not may come down to both the female's desirability and the social dominance status of the males. Stable male hierarchies with a powerful alpha in charge usually produce relatively peaceful relations among the males when it comes to mating. During periods of dominance upheaval when no male is firmly in control, or when two or more are vying for that control, a female coming into her swelling cycle can create absolute pandemonium.[22]

At Gombe, when an alpha male headed off to some distant valley with a female in tow, the Tanzanian assistants would speak of them being on safari (the word *safari* meaning "a journey" in Kiswahili). It implied a mutual interest in being alone together, but such is not necessarily the case. While most mating takes place in the group opportunistic setting—at Mahale the figure is more than 80 percent[23]—males also employ a consort strategy. A male "persuades" a female to accompany him away from the rest of the community into a distant area, often near the periphery of the community's territory. The persuasion can be coercive and may involve the same signals that males give to indicate desire to mate, done repeatedly over many hours. The female may or may not go willingly, and if she is already swollen, the interest from other males may also interfere with the lothario's scheme to take her away. If she doesn't respond to these actions, the male may be more overtly aggressive toward her. While the female is usually in or nearing the swollen portion of her cycle, males also try to take nonswollen females off on consortships, and sometimes these females spend the duration of at least one full potential conception cycle away from the community in the company of just one male. According to Goodall, males are often most aggressive during the early hours and days of the consortship, since this is the time when females are most prone to try to escape.[24]

If this mating strategy sounds a lot like a kidnapping, it does sometimes resemble it. Consortships can last from a few days to more than a week. But females are active, not passive, in the courtship and mating process, and they may slip away at any point in the consortship. The pair often ends up in the vicinity of the neighboring community's territorial border, which is dangerous ground. Both male and female behave warily due to the risk of running into an enemy patrol. Consortships are not necessarily a highly effective reproductive strategy; in Gombe, for example, only 14 of 117 consortships (12 percent) are believed to have resulted in pregnancies.[25]

Do Males Choose Too?

A pendulum in primatology has swung between a focus on males and a focus on females as the active choosers of mates. In the early decades of research, little attention was paid to female mate choice. Males were more easily accustomed to observers, were more easily observed, and exhibited "more interesting" social behavior from the perspective of the male scientists who carried out the field research. Following a tidal shift in the 1960s and 1970s, female choice received most of the attention. Only in the past decade and a half has male choice become a prime research topic again. Males do exert mate choice. Male baboons prefer females with large swellings, and, as we saw, captive male chimpanzees appear to do the same. Martin Muller and his colleagues Emery Thompson and Richard Wrangham showed that male chimpanzees mate preferentially with older females. The common stereotype that men prefer younger mates is not just a stereotype; cross-cultural studies have shown that men tend to choose younger partners, just as women typically choose older ones. But among our closest kin the pattern is the opposite. Among the Kanyawara chimpanzees, males had a clear-cut preference for older females. They approached old females and made solicitation gestures more often. They congregated most often in parties containing swollen females when those females were among the oldest in the community. The older the female, the more likely she was to mate with the alpha and other high-ranking males.[26]

Why would male chimpanzees actively prefer old females to younger ones? It's not because the younger ones were not mature enough to produce offspring; older mothers were also preferred to young mothers. Muller and colleagues argue that key life history differences between humans and chimpanzees account for the preference for older females among chimpanzees. Menopause is an abrupt and complete cessation of fertility in women, which does not occur in female chimps. Spending energy and time associating with and mating with a thirty-five-year-old female chimpanzee is therefore as likely to increase a male's paternity chances as with a twenty-year-old female. The reproductive value of a female chimpanzee, from a male perspective, is long term compared to that of a human female. A male only attempts to control a female while she is fertile and doesn't bother with forging long-term partners. At the same time, she remains fertile to nearly the end of her life span, very unlike a human female. A potentially confounding factor that Muller's study doesn't address is that old females are more likely to be high ranking than young females; perhaps males are attracted first and foremost to social dominance. This would make eminent sense since high-ranking females enjoy high reproductive success. But it's still an intriguing finding, particularly since all the visual cues that human males obsess over, from youthful bodies to wrinkle-free faces, apparently play no role whatsoever in mate attraction in chimpanzees.

Males exert mate choice indirectly by restricting the ability of females to freely choose with whom they mate. At times they try to prevent females from mating with other males simply by guarding them. At Ngogo, Watts described mate guarding by coalitions of males who cooperated aggressively to keep females away from other potential suitors. These coalitions may have formed because there are so many males in the Ngogo community that no one male, not even the alpha, could monopolize a female. Two- and three-male coalitions enjoyed greater mating frequency for each participating male than any single male guarding a female obtained when a female was swollen, and they achieved this with a lower degree of aggression. Notably, these mate-guarding coalitions form based primarily on the mutual benefits to the allies of controlling mates, not because they are necessarily brothers or cousins.[27]

Males sometimes take a far more sinister approach to restricting female choice than simply guarding her access to rival males. Sexual coercion and sexual violence are commonplace among chimpanzees. Sexual coercion is found across the primate order. Among chimpanzees, it is a routine part of male social behavior everywhere the species has been observed. But since females become more selective in the choice of males around the time of ovulation, the presumption has always been that male coercion must not be very effective. Muller and his collaborators disagreed. They pored over more than a decade of field observations from Kanyawara and discovered that males who are highly aggressive toward females achieve more matings with those females. They also found that when females have sexual swellings, they solicit matings more from males who have been aggressive to them throughout their cycle. This pattern was irrespective of a male's dominance rank; even lowly males beat up on females in hopes of forcing them to cower into submission, and sex.[28]

The pattern is hardly limited to Kanyawara. Over a seventeen-year period, Gombe male aggressiveness toward females during their swelling cycles was highly associated with how often they mated with those females. Most noteworthy was the finding that males who were aggressive toward females throughout their *nonswollen* periods were most likely to be the fathers of their offspring. In other words, long-term male coercion and intimidation work, and they are likely an evolved adaptive strategy. Given that we look to chimpanzees for many insights about the origins of our own social behavior, this is a highly provocative finding.[29]

It's important to note that Christophe Boesch and his team working in Taï National Park on the other side of the African continent have never reported sexual coercion among their chimpanzees. Taï females are therefore presumably better able to be choosy about male partners without the threat of aggression from them.[30] Joseph Feldblum and his coauthors, who conducted the Gombe analysis, argue that because there are fewer males at Taï and they are more gregarious than at other sites, a male's need to use aggression to get the mate he wants may therefore be lower.[31] Although there may be local exceptions due to the particular demographics of some chimpanzee communities, a female's options in choosing a mate may be more constrained than they appear.

Migration Decisions

Female chimpanzees emigrate at or after sexual maturity and eventually settle into a nearby community for the remainder of their adult lives. In most chimpanzee studies, all or nearly all of the adult females in a community are immigrants. Some of these are fairly new arrivals and, being unaccustomed to human observers, they remain shy for months or years. Others may be longtime residents of the community but remain peripheral and travel mainly in the outer areas of the community territory, or perhaps in the overlapping territories of more than one community. They may be low ranking compared to the females who occupy the core area of the community's territory. It is sometimes hard to know the exact number of females in the community, because peripheral members may move back and forth between their home community and their adopted one. For all we know, some may be residents of multiple communities simultaneously. The lives of these peripheral females remain a bit mysterious because, since they are encountered less often by researchers, they remain wary of people and don't allow close observation. Understanding their life histories better would inform us about the reproductive ecology of females more fully, but that won't be easy information to obtain.

During late adolescence, females settle permanently into another community. Unlike female gorillas, they rarely transfer again after that initial migration, perhaps because if they did they would likely have an infant in tow who might not be welcomed by the members of the new community. This migration occurs in all social mammals, from whales to primates. Migrant females normally arrive alone, but occasionally a female arrives with an infant and that infant may become an integral member of the community. During my work at Gombe, Beethoven was a low-ranking adult male who had arrived as an infant with a young immigrant female who was likely Beethoven's older sister. Emery Thompson and her collaborators reported a possible case of the immigration of a number of females who were accompanied by their young offspring into the Sonso community at Budongo.[32] But paternity analyses by Kevin Langergraber and his coauthors showed that these male offspring of the putative immigrant females were actually fathered by Sonso community males. The "immigrant" adult females who were thought to be bringing their sons into a new community with them were in fact residents of the Sonso community that were just shy or peripheral enough that they could be mistaken for new immigrants.[33]

The immediate cause of emigration of males in those species in which males emigrate is usually increased aggression from an older male (perhaps the father) who begins to see the young male as a threat. The immediate impetus for female chimpanzees to migrate is less clear. The deeper evolved reason is, we believe, to avoid inbreeding. The benefits of transferring must outweigh the costs of leaving an area that she knows intimately. She's also forsaking the company of chimpanzees whose web of relationships is well established in favor of the unknown. While most females rise in rank throughout their adult lives, with concomitant access to desired males, meat from hunting, and other perks, some immigrants remain lowly their entire lives. Migration is full of uncertainties for the female.

The female's arrival in a new community is received differently by the resident males and the resident females. To the males, she represents a new opportunity to enhance their reproductive success, and they welcome her with open arms. If she arrives with a sexual swelling, she will use it to smooth her passage into the embrace of the new males. Even if she immigrates with an infant, the males will accept her readily, quite unlike nonresident females with infants that are encountered during male patrols, who are attacked savagely. But to the resident females, the immigrant's arrival is not good news. Resident females often display aggression toward the new arrival, who is not only vying for the attention of the males but will be seeking out the same fruit sources as the residents. Sonya Kahlenberg of Harvard University and her colleagues examined the issues faced by immigrant female chimps. Resident females may outright attack immigrants, injuring them, and, if they had the temerity to arrive with a baby, attack the infant as well. The immigrants incur higher stress levels as a result of aggression from the resident females, judging by their increased levels of urinary cortisols. Males of the new community frequently intervened to stop resident females from harassing immigrants, and they were often effective in mitigating aggression toward them. The immigrants took advantage of male protectiveness by spending time with them, even outside their swelling cycles, and this proximity helped in reducing aggression from other females.[34] Similar observations have been made at Gombe, Mahale, Budongo, and Taï. In the Sonso community at Budongo, males were less able to protect females after a number of females immigrated at roughly the same time. Researchers there believed the males were ineffective protectors simply because they were outnumbered by resident females.[35]

Not every female migrates at maturity. At Gombe, some members of the famed F matriline—Flo and her descendants—have remained in the Kasekela community throughout their lives. When Goodall arrived in 1960, Flo was already an older female, so we don't know whether she had immigrated. But her daughter Fifi, who bore nine offspring and lived to a ripe age, never emigrated, and neither did one of her daughters. Why females of this lineage, as well as occasional other females at Gombe and other study sites, don't necessarily migrate is unknown. Speculation has always been that because the F lineage tends to be high ranking and socially powerful, there are incentives for staying that outweigh the presumed inbreeding-avoidance rationale for transferring. Females who stay also reap the benefit of inheriting a range that they know well and that may be located in prime real estate compared to where they would settle elsewhere. If no females migrated, inbreeding and genetic issues would eventually ensue. But perhaps some limited, frequency-dependent failures to migrate are actually reproductively advantageous for well-positioned females.

Predicting exactly when and how a female will transfer to another community is difficult. Stumpf and her collaborators examined the question among Kanyawara chimpanzees, and the results were somewhat surprising. At Kanyawara, a decade of data showed that several factors played important roles in the pattern and timing of female migration. Stumpf found that females transferred at an average age of twelve and a half to thirteen. Most transferred while fully swollen. And most were observed mating with at least one adult male of the new community during that first swelling, meaning they were sexually active at or shortly after the time they arrived in their new home. In fact, the time to first observed mating after her first swelling was much shorter (two months compared to seven) for immigrant versus females born in that community. Immigrant females also ended up giving birth sooner compared to the one natal female who did not transfer (twenty-seven months versus thirty-seven months after their first swelling). Surprisingly, natal and immigrant females mated equally often with the males of the community. The natal females could have been mating with their fathers. As we saw earlier, there was no indication that social stress, based on behavioral and hormonal measures, was the immediate cause of a female leaving the community of her birth.[36]

In other words, the trigger for migration in Stumpf's study was not the usually invoked reasons of aggression or stress. We could assume inbreeding avoidance, although most maturing Kanyawara females were already mating before they migrated. As at other study sites, females did not appear to necessarily avoid sex with all males who might have been their uncles or cousins. The fact that most immigrating females were swollen lends support to the idea that the swelling is a "passport," as Boesch has dubbed it, providing a new female with a relatively smooth entry to the new group of males she will be living with for decades to come. Two females in the study transferred without swellings, however, and no observations supported the idea that these two faced undue aggression from either males or females as a result.

The Kanyawara study concluded that diet quality might be an unappreciated cause of migration. Females tended to transfer between communities during periods of fruit abundance and high fruit consumption. This suggests that females wait until they are well nourished before undertaking the risky journey into a new community, where food resources would be initially difficult to find and of uncertain abundance.

A female who arrives in a new community brings her own tool using and other traditions with her. One might expect that her fellow community members would observe and adopt these new behaviors, but this does not necessarily happen. Lydia Luncz and Boesch of the Max Planck Institute for Evolutionary Anthropology examined cultural diversity across communities in relation to female migration patterns. They found that males and females in a community varied very little in cultural traditions despite the fact that the females had grown up learning traditions of a different community. Nut-cracking tools did not vary between those females and males over a twenty-five-year period. Females who immigrated adopted the behaviors of their new community instead of clinging to old traditions, which might then have been observed and adopted by new group-mates. Luncz and Boesch believe that because females transfer at sexual maturity and most cultural innovation occurs among younger animals, by the time the female arrives in the new community she is not highly disposed to learn new behaviors. The social pressure of learning the social web of her new community is an inducement to make small shifts in her cultural behavior to adapt to her new surroundings.[37]

The Decline

Unlike human menopause, in which a woman's ovulation cycles end abruptly in middle age, fertility in female chimpanzees declines with advanced age but does not cease. Emery Thompson showed that, among

Gombe chimps, reproductive output peaks between ages fifteen and thirty and does not begin to show marked decline until past age forty (by which time many females in the wild have already died). Even then, healthy females continue to have babies at the average rate of the population well into their forties. After that, fertility declines and menstrual cycles become somewhat less regular, although cycling shows no marked decline until the very end of a female's life.[38] A study by Sylvia Atsalis Northeastern Illinois University and Elaine Videan of the University of Michigan compared wild and captive female chimpanzee reproductive aging. They found that captive females underwent something resembling menopause and suggested it might be because captives have their first cycle at a younger age, then cycle more regularly than wild females, presumably due to better nutrition. Captive females live to far more advanced ages overall than their wild counterparts, thereby surviving to the point of egg depletion, which wild chimpanzees never reach.[39]

Emery Thompson suggests that, for female chimpanzees, reproductive aging is simply an aspect of overall aging, and not a thing apart from it as it is in our own species. Egg depletion has been estimated to take place at around age fifty in captive female chimpanzees, which is more or less on the human schedule.[40] James Jones of Stanford University and his coauthors showed that, as females age, they often rise in dominance status, which may in turn enhance their reproductive rate and their infant survival odds.[41] This again presents a very different schedule of reproductive value and associated life history factors from that seen in any human population.

Kristen Hawkes and Ken Smith of the University of Utah compared lifetime fertility curves for humans (in this case, birth records from the 1800s for a particularly fertile group—Mormon settlers) and wild chimpanzees. They found both similarities and stark contrasts. The average age of a female at the birth of her last child is quite similar between humans and wild chimpanzees—forty or so. This is despite the fact that women live a long postreproductive life. Great apes, meanwhile, continue their high fertility levels right through old age. Our two species' maximum recorded life spans differ by nearly 100 percent: 122 years for our species versus 66 for a chimpanzee. Although the average life expectancy in many human populations in developing countries is still only fiftyish, there are still plenty of individuals who, having survived the perilous early years, live to great ages.[42]

Human populations have a highly bell-shaped age-specific fertility curve. Among Mormon settlers, with no confounding effects of contraception, no one conceived before about age fifteen, followed by a steep peak in fertility at every age through the midforties, at which point fertility fell quickly to zero. Age-specific chimpanzee fertility shows a far flatter curve, with the rate reached by age twenty remaining more or less the same well into the forties. Emery Thompson and colleagues showed much the same results in a comparison between wild female chimpanzees and two groups of traditional people, the !Kung of southwestern Africa and the Aché of Paraguay. It may be, as Emery Thompson and colleagues suggested, that the highest-fertility female chimpanzees are the ones who live longest, because their fertility may be linked to social dominance, nutrition, and other factors conducive to long life. If low-fertility females are more likely to die young, it would give the appearance of very high fertility to older females.[43]

We can say with certainty that wild female chimpanzees rarely outlive the fertile period in their lives, while human females normally do. In many traditional human populations—the so-called hunter-gatherers—fully a third of women are over age forty-five and so likely no longer bearing children. Hawkes resolved the evolutionary paradox of menopause with her "grandmother hypothesis."[44] Great ape babies are in the care of their mothers, and if a mother dies, her baby will almost certainly die in rapid order. While the mother is struggling to care for her infant, she receives no direct help from any other member of the community. In our own species, females stop reproducing at a relatively much earlier point of their lives than our ape relatives do. Postreproductive women then turn their attention to assisting their daughters in offspring rearing. They help their daughters gather and carry food and they care for small children to allow their daughters to do other work, or simply to recuperate a bit from the energy stress of mothering. For a species that does not simply acquire food hand to mouth but rather gathers, carries, and stores it, a grandmother's help could be a very big deal.

The benefit of the grandmother's reproductive success, and of that of her daughters, who share 50 percent of her genes, would be a greater overall number of surviving children than either would have had if the grandmother herself continued to reproduce. This part of the model has been challenged in recent years. The grandmother hypothesis assumes that the common ancestor of humans and great apes had an apelike life span and reproductive trajectory, and that humans evolved a long life extension postmenopause. Although some have argued that female chimpanzees do experience a menopause of sorts, Emery Thompson and others emphatically refute that claim.[45] It's clear that in the course of human evolution, natural selection favored an extended life span but not a concomitant extended reproductive life span.

Human Sexuality and Recent Chimpanzee Research

This chapter may give the impression that sex in chimpanzees is a free-for-all in which male might makes right when it comes to access to females, and therefore reproductive success. But that may not be exactly the case. When we try to parse out the chimpanzee mating system, male reproductive skew—the distribution of matings and offspring conceived among the adult males—is an important but misunderstood consideration. We know that alphas and a few high-ranking males dominate conceptions in most communities. Reproductive skew theory predicts that high-ranking males will allow low-ranking males to mate on occasion, since they need their help in other tasks, such as patrolling the community's territorial borders. Or perhaps the alpha is simply unable to control all those lower-ranking males, who end up getting the mating "leftovers." We will see that the reproductive success of the alpha male decreases as the number of competitors and synchronously receptive females increases.

Recent evidence suggests that chimpanzee societies are not as starkly different from human mating systems as one might think. They live in a fluid society made up of temporary subgroups and transient mating bonds between males and females. But underlying that system is the strength of the bonds between particular male-female pairs. Even in a huge community like Ngogo, there are long-term bonds among particular males and females. Langergraber and his coauthors showed that males of all ranks maintain affiliations with females that are based on something other than kinship or nepotism. The authors speculate that because male chimpanzees tend to inherit the home ranges of their mothers, they form lifelong bonds with other females that occupy the same areas of forest. These associations translate into reproductive success for those males, because they end up mating preferentially with them.[46] This effect is not limited to chimpanzees; important bonds among opposite-sex group-mates that result in mating have been found in other social primates such as baboons as well.

It has long been supposed that long-lasting human pair bonds are the result of the necessity of parental care by both mother and father to rear an infant. In Hrdy's words, we're all basically cooperative breeding units. Chimpanzees, utterly lacking in paternal care, don't fit that notion at all, nor do the other great apes. But it's beginning to appear that the origins of a long-lasting pair bond might, at least in some form, be nascent in chimpanzee society.

CHAPTER 6

Growing Up Chimpanzee

AFTER A PREGNANCY that averages eight months, a chimpanzee mother gives birth to an infant that is nearly as helpless as a human newborn. For the next several years the infant will be utterly dependent on its mother and, for several years beyond that, it will need the psychological support system that only the mother can provide. Critical developmental hallmarks happen during these first several years, and even after weaning, juvenile chimpanzees that lose their mothers are marked throughout their lives by stunted size, low social status, or both.

Like humans, chimpanzees give birth to a single offspring. But recent research has confirmed what most chimpanzee researchers had long observed: they are more likely to give birth to twins than we are. John Ely of the Alamogordo Primate Facility and colleagues compiled twinning data from captivity and found that the monozygotic—identical—twinning rate of one pair every 230 births is about the same as in humans. But the rate of producing dizygotic—nonidentical—twins is once every forty-three births, more than twice that of women who are not using fertility-enhancing drugs. Female chimpanzees who have twins are *five* times more likely to give birth to twins again. The same pattern holds for women; twinning has a genetic basis, and the odds of twins increase with maternal age.[1] Paternal effects on twinning are poorly understood. There are rare human cases of dizygotic twins produced by two different fathers. It's tempting to hypothesize that dizygotic twinning might be adaptive in a species whose females are as promiscuous as chimpanzees are. A female who mates with a dozen or more males in a day while ovulating would potentially be offering paternity to two males, which might enhance her ability to lessen aggression from all the males in the community later.

Birth is less of an ordeal to a mother chimpanzee than a woman, because the size ratio between the widest part of the newborn's skull and the narrowest stretch of the mother's birth canal is not as tight. The chimpanzee skull, one-third the size of that of a human infant, passes through hips that have not evolved the shorter, broader dimensions that our own bipedal posture requires. There is enough space between the baby's head and shoulders and the mother's body that she can give birth unaided. Most first-time human mothers in traditional societies get through the process only with the help of a midwife or two. The very few Western scientists who have witnessed a wild chimpanzee birth have described a mother in obvious but not disabling discomfort who delivers the newborn into her own waiting hands. She bites through the umbilicus, delivers the placenta, and either eats it—sometimes sharing with other members of the community—or pushes it under some leaves, and the birth is done.[2]

For the next couple of months, the infant is inseparable from the mother's abdomen as she travels the forest with her baby clinging to her. At around three months, the infant begins to emerge from the cocoon of its mother's arms, approaching other infants to play under her watchful eye. Socializing with peers increases in frequency and intensity over the next few years. But in those earliest months and years, the vast majority of interactions the baby has with the world are with its mother and her milk supply.

After fifty years watching the mother-infant bond at the center of chimpanzee life and the importance of nursing to it, researchers finally began to examine the mother's milk. The milk of all the great apes has a lower fat composition than human milk, 2 percent compared to nearly 4 percent. It is also slightly lower in protein. Katie Hinde of Arizona State University and Laura Milligan of the University of California, Berkeley, surveyed milk and lactation across the Primate order and showed that stress hormone levels in a mother's milk could be inferred from the female's life history. The levels also predicted aspects of her infant's temperament. Instead of viewing milk in isolation, they considered the roles of both milk production and nursing in evolutionary context. Many marine mammal mothers, for example, produce milk that is incredibly high in fat—greater than 50 percent, to the point that it has nearly the viscosity of ketchup—because the cold marine environment requires rapid deposition of body fat in infant development. Species in which lactation occurs infrequently, because babies are hidden in a shelter for a day or more between nursing opportunities, also produce highly concentrated milk nutrients. Animals in hot, arid environments often produce milk highly diluted with water, probably to keep the infant hydrated. With these adaptive suites in mind, Hinde and Milligan examined where chimpanzees fit into the range of nonhuman primate milk. They found that mountain gorillas, who live on a highly leafy diet, produce milk that is quite protein rich compared to that of frugivorous chimpanzees and bonobos. They also tested the long-standing notion that big primate brains should place a premium on nutritious milk, although which nutrients are most critical for growing a big brain has been a subject of debate.[3] The greatest caloric value in chimpanzee milk is from fat. But some researchers have argued that glucose in the body is badly needed by the brain for metabolism, so a sugar-rich, high-calorie milk might be desirable. Chimpanzee milk (as well as that of humans) is overall low in fat and in energy density and only moderately rich in sugar. Sugar and fat are almost always in inverse proportions to each other in animal milk. Based on studies of mother baboons, as maternal health or weight declines, the quality of the milk appears to hold steady, but the quantity slips.

Theories of sex ratio evolution posit that there may be an advantage to mothers who invest more time and energy in infants of one sex versus the other. A mother who is healthy and socially dominant might produce more sons than daughters if those sons inherit her high status and are therefore able to mate prolifically and spread her genes far and wide. A female will not lack for male suitors even if she isn't highly ranked, so low-status mothers might be expected to produce more daughters. One might expect mothers with sons to produce the highest-energy milk possible to get males well started on the road to robust body size and high rank. Mothers of females would produce milk that is lower in fat and protein. Captive studies of rhesus macaques have in fact shown this to be the case. A confounding factor is that sugar content changes the osmotic potential of milk, altering its balance with fat as a result of more water being drawn into the breast tissue. In humans as well, sons get more energy-dense milk and grow faster and heavier than female babies, though Hinde and Milligan point out that this may be due to differences in how milk of various energy compositions is digested and not its overall nutritional value.

There is some support from field research for at least one of the ideas presented here. At Taï, Christophe Boesch found that high-ranking mothers invest more in their sons than their daughters. The interbirth interval, when controlling for early deaths and disappearances, was more than twice as long when a dominant mother was raising a son. The sons of high-ranking females also had higher survival rates to adulthood than other offspring. So dominant females' investment paid off in the survival odds of their sons. Socially dominant mothers invested about the same amount of time raising daughters as lower-ranking females invested in raising their sons. Remember that male chimpanzees will remain in the community for life, while female offspring will likely emigrate. But this trend toward favoring sons was not found in Gombe, which Boesch argues is due to the tendency for some Gombe females to stay in their birth community instead of migrating to another. There may even be a supportive relationship among the mothers and daughters who don't transfer, providing some countervailing incentives for females to ignore whatever inbreeding-avoidance imperative exists.[4]

Beyond maternal investment in offspring is the question of whether fathers can help their own offspring. Because the chimpanzee mating system is so promiscuous, one would think not. But Julia Lehmann and her colleagues at the Max Planck Institute for Evolutionary Anthropology argue that there is circumstantial evidence at Taï that males recognize their offspring using association cues and treat them differently. Based on DNA paternity evidence, Taï males did not show any association preference for females with whom they'd had babies, but they were less aggressive toward females with newborn infants in general. This wouldn't necessarily mean much, since many males likely mated with that female and each had a small chance of being the father. But these fathers continued to be less aggressive to those offspring long after males who had not fathered babies returned to their normal baseline levels of aggression. And males, who seldom engage in social play, spent more time playing with their own offspring.[5] So even though there is no paternal care and males cannot know with any certainty which babies they've fathered, they may use indirect cues based on their relationships with mothers to treat their own progeny favorably. This is certainly paradoxical given the nature of chimpanzee society and the fact that even in human societies, men don't always know which children they've fathered and may hold notions about sex and paternity that don't reflect biological reality.

We have known since the earliest days of Jane Goodall's work that infant development in chimpanzees parallels many aspects of human infancy. Mothers devote years of energy and resources (milk) to raising a healthy baby. A recent analysis by Melissa Emery Thompson and colleagues indicated that offspring that are weaned early end up with smaller body size as adults. Body size is a predictor of infant survival in many animals, and adult body size can be a predictor of mating success. A female who cuts her milk production early to produce more offspring may therefore be compromising the future success of her offspring.[6]

Differences between males and females of many animal species appear early in life, and chimpanzees are no exception. Elizabeth Lonsdorf of Franklin and Marshall College and her colleagues compiled developmental differences through the early years of life for males and females, and striking differences emerged. Males and females spent equal amounts of time nursing in early infancy. Males began to travel independently of the mother earlier and more often than infant females did, which corresponds to similar findings in human infant development. As infants begin to travel independently from the mother, usually walking behind or next to her, they remain in contact and communication with the mother.[7] Marlen Fröhlich of the Max Planck Institute for Ornithology in Seewiesen, Germany, and colleagues recently reported that while traveling through the forest, infants and their mothers coordinated their movements through gestures and vocalizations. Mothers modulated their calls to the ability of their infants to receive them. There was, in the authors' words, an ongoing social negotiation.[8] Frans Plooij had studied play and the relationship between mothers and their infants at Gombe in the early 1970s.[9] Frölich's study provided an updated view. As one would expect, mothers initiated travel when they set off with their infants, and they used gestures to do so. This body language varied among females within a community, and also across communities. Infants responded with a repertoire of whimpers and other calls, transitioning to a combination of calls and gestures as they grew older.

Lonsdorf and her collaborators found that young females acquire basic tool-using proficiency at Gombe up to two years earlier than males do. Males, meanwhile, had more interactions with more social partners than females did around the time that both sexes begin to separate themselves physically from the mother. It is at this time—when the male is around two and a half years old—that he begins to make connections to the other members of the community.[10] These new social partners were mostly other male infants, and, on average, they remained more sociable than did females. It wasn't entirely clear whether the mother played a role in selecting the social partners her infant engaged with. Carson Murray and her colleagues found that mothers who were rearing sons were more sociable with the rest of the community than were mothers who were rearing daughters, especially in the early months of the infant's life.[11] This suggests that the mothers in Lonsdorf's study were making decisions about how and when their offspring engage with their social environment, which had a potentially major impact on sex role development of the offspring.

Play

Play is an odd concept in the study of animal behavior. There is solitary play—manipulating an object or frolicking—which not only mammals and birds but even some reptiles do. Social play is more characteristic of higher animals and is not always easy to define. It's immediately obvious when two puppies or baby monkeys are playing, and yet describing play unambiguously is surprisingly difficult. Social play often looks like aggression and consists in large part of exaggerated movements, postures, and gestures, many of which are also seen in fighting. Play is usually seen as an early opportunity to rehearse skills needed later in life, without any of the associated risks: play fighting, play mounting, play mothering, and so on. In a species as social as chimpanzees or gorillas, when infants are present, one will see plenty of play behavior. This leads us to believe that play is biologically important. But there are many species that live in pairs or in which females are solitary, so that infants grow up with few opportunities for social play with peers. And yet there is no evidence that such play deprivation affects the psychological well-being or social development of those play-deprived individuals.

Matthew Heintz recently analyzed long-term data on social play from Gombe in order to understand the role it plays in later life. He found social play to be equally important in the lives of both male and female infants. As infants developed better motor skills, the intensity and complexity of their play increased. He also found that stress hormone levels increased with play, an unexpected finding because past research had indicated that play functioned as a stress reducer. Heintz's goal was to identify long-term benefits or associations later in a chimpanzee's life that might be predicted by play patterns in infancy. But there appeared to be none. Even though we see infant males dominating each other in play, infants that were playful didn't rise in dominance rank more rapidly as young adults. They didn't produce offspring any earlier than their less playful group mates. The most playful infants did not become the most social adults or the most groomed adults.[12] We understand many of the immediate benefits of play. But how infant play might translate into reproductive success or long-term survival, as has been shown in other mammals from rodents to bears, remains essentially unknown. If play is adaptive—and the amount of time and energy infants invest in it suggests it must be—it isn't entirely clear how.

Lonsdorf and her colleagues found that social play peaks at two years old, and that males engage in social play more often at an earlier age. Females ultimately become more dexterous tool users than males ever do, and they tend to copy their mothers in ways that males don't.[13]

Orphaned

One of the best-known anecdotes from Goodall's early years at Gombe was her account of perhaps the most famous orphaned animal in history. Flo was a powerful and aggressive female who was estimated to be well into her forties at the start of Goodall's work at Gombe. Venerable Gombe chimps Figan, Faben, and Fifi were among her older offspring. But by the time she gave birth to her last infant, she was elderly, and her mothering skills had suffered. Her last infant, Flame, died when her next-to-last offspring, Flint, was five years old. At that point Flint became somewhat infantilized again, his bond to Flo becoming stronger rather than weaker as he grew up. When Flo died a few years later, Flint was nearly nine years old and approaching adolescence. But newly orphaned Flint was not able to cope with life following the loss of his mother. He stopped eating, became listless, and exhibited the signs of what appeared to be clinical depression. Only weeks later, Flint died of unknown causes that were likely related to the psychological trauma of the loss of his mother.[14]

Breaking the bond between mother and infant creates a disruption in the infant's life that can range from troubling to traumatic. Not all orphans suffer Flint's fate, but most suffer to some extent for the rest of their lives. Maria Botero of Sam Houston State University and her colleagues documented clear anxiety-related behaviors among Gombe orphans.[15] Elfriede Kalcher-Sommersguter of the University of Graz, Austria, and colleagues examined the lives of captive chimpanzees that had lost or been taken from their mothers at an early age versus those that had not. They found that maternally deprived chimpanzees suffered lifelong consequences regardless of their social environment later in life. They groomed less, socialized less, and were less socially integrated than chimpanzees that had mothers early in life.[16] Michio Nakamura of Kyoto University and his colleagues reported that orphaned male chimpanzees tend to have shorter life spans than their group-mates even when they survive the trauma of being orphaned.[17] Perhaps because of the male philopatric social system, in which adult males remain in contact with their mothers throughout life, losing one's mother even as a young adult is an isolating experience that has long-term ramifications.

Orphans tend to fare poorly in chimpanzee society, mainly because adoption is very rare. Compared with other nonhuman primates, allomothering—the care of an infant by a female other than the mother—is quite rare. Toshisada Nishida found that among Mahale chimps, around 3 percent of the time of an infant under two years old is spent in the care of a nonmother female.[18] This lack of connection to other females in the community leaves infants stranded when the mother dies. In some primate species, an orphaned infant or juvenile might find a caregiver, or at least a protector, who would give the orphan a chance in life. But adopting an unrelated infant is an act of extreme altruism. In a recent survey of the fate of orphans in Taï, Boesch and his coauthors reported that of thirty-six infants and juveniles orphaned over a period of twenty-seven years who survived at least two months, eighteen were adopted. This does not include small infants who lost mothers and disappeared either with them or shortly after. Orphans less than five years old did not survive the loss of their mother. Juveniles older than five who survived were often greatly stunted in physical and social development. Sometimes an adult may offer short- or long-term social support, usually in the form of staying in the company of the orphan or waiting for the orphan to catch up when the adult sets off on a trip.[19]

Eight Taï adoptions were made by females. One was by an older sister and three were by females who had close bonds with the dead mother. But most surprising was that the majority of orphans were adopted by adult males. Six adult males and three older brothers performed ten of the adoptions. These males shared food, including meat, with their adopted orphans; waited for them while traveling; and supported them in conflicts. In some of the cases, the orphans were young enough that they needed to be carried, and they rode on the backs of their adoptive fathers or older brothers. In only one case was the adoptive parent the known biological father of the orphan. These are examples of perhaps the most altruistic behavior that wild chimpanzees are known to perform. They demonstrate that chimps are sometimes empathetic to the needs of unrelated members of the community. Boesch and his collaborators argue that a male who adopts an infant may do so with the expectation that the orphan will grow up to be his ally in conflicts within or between communities. But only half the adoptions were of male infants; the female orphans who were saved by adoptive fathers or brothers would simply emigrate at puberty. One would assume that the survival of adopted orphans would be higher than that of orphans who are not adopted, but the study was unable to demonstrate this due to high overall mortality in the community.[20]

Catherine Hobaiter of the University of St. Andrews and her colleagues surveyed adoption patterns in the Sonso community at Budongo. They documented eleven orphans over a period of twenty-one years, seven of whom were adopted. Of these seven, all but one were cared for by older, independent siblings born to the same mother. Orphans adopted by an older sibling had a significantly higher likelihood of surviving to puberty than those orphans who were not adopted. Comparing the fate of all orphans in Budongo, Taï, Gombe, and Mahale, survival to one year after loss of the mother was 80 percent for orphans adopted by other adults. It was 100 percent for those adopted by adult or subadult older siblings, and 70 percent for orphans adopted by immature siblings. Adoption is thus done mainly by older siblings. Unrelated adults take on the task only when older siblings either are not present or fail to adopt. Sibling adoption can be an effective survival opportunity for orphaned young chimpanzees.[21]

Almost Grown

The life stage between utter maternal dependency—infancy—and reproductive maturity is the juvenile period. Juveniles no longer need their mothers to survive, nor do they fit into the dominance hierarchy of either sex. Chimpanzees have a protracted life cycle compared to most other primates, which, along with the energetic constraints of development, allows for the learning and socialization that are so critical to becoming a chimpanzee. There's an inaccurate perception that juveniles lead a fairly carefree life. They've survived infancy but have years before they begin to take on adult roles and all the stresses that come with adulthood. But the years between late immaturity and early adulthood are just as awkward a time for chimpanzees as they are for humans. Once they reach full physical independence, juvenile chimpanzees between about five and eight years old are too old to nurse, and their mothers have shifted their attention to the next infants. This leaves the juveniles on their own, often struggling to keep up with foraging parties as they travel through the forest. Indeed, Herman Pontzer and Richard Wrangham showed that at Kanyawara, juveniles slow the rate of travel of their mothers, who would otherwise keep up with group-mates even when carrying an infant. Such limitations on ranging distances may impact a young juvenile's ability to reach the food trees it desires, and it may also limit the mother's ability to do so. Only when the juvenile approaches ten years old is he or she traveling at the speed of the rest of the community.[22]

Over the past decade, our view of growing up as a primate has changed. Once seen as purely transitional stages between infancy and adulthood, the juvenile and adolescent stages of life history are now understood to have a trajectory of their own and are under evolutionary pressures unique to that period of life. Of course, we've known this is the case in our own species for a very long time; consider the attention paid to the issues of both adolescent girls and boys in society. The same is true for the great apes.

Between ages five and ten, chimpanzees gradually separate themselves from their mothers. Male juveniles tend to gravitate toward the company of adult males, and later adult females. The mother begins cycling and consorting with various males years earlier, which creates the first real wedge between them. By the age of ten, females begin visiting other communities for weeks or months at a time. Male juveniles approach puberty and begin to associate more with parties of adult and subadult males than with their mothers. There's a distinct loneliness to being a juvenile, prepubescent male chimpanzee. Once the separation from the mother is complete but the male is not yet integrated into the company of the older males, the older juvenile and preadolescent males become hangers-on. They spend most of their time with other juveniles but also try to travel with male parties and patrols. Their presence is tolerated, although they receive little reciprocated grooming or support in conflicts. Their main grooming partner remains the mother. And they also spend a good deal of time alone. Females, meanwhile, spend much time with unrelated older females and other juveniles.[23]

Preadolescence in females is marked by the first sexual swelling, at around age eight. These increase in size and regularity over the following couple of years, and the females begin to mate with males by age eleven. Behaviors of youth such as social play and socializing with infants rapidly diminish. This newfound maturity elevates the young female from a hanger-on at social gatherings to an accepted community member whose company is sought out by males. By the time she is thirteen and potentially fertile, her presence is welcomed by adult males after, for instance, a successful hunt when meat is being doled out.

Male chimpanzees, like human males, undergo the physiological changes of adolescence long before they begin behaving like adults. They begin to produce sperm before the age of ten but likely won't be mating with fertile females until years later. Their participation in foraging, patrolling, and hunting parties is limited. Early adolescent males mate with adult females, usually out of sight of older males. They even try to cajole females into accompanying them on consortships, usually without success. Only by late adolescence do they have the physical size and strength to stand tall with adult males and to coerce females to join them on consortships. The adult males become less tolerant of an adolescent male who tries to assert himself in their presence. As adolescent males begin their rise in the dominance hierarchy, they first challenge each adult female. By the time adolescence is merging into adulthood, a young male has likely raised his social status above that of all the females and is beginning to challenge the lowest-ranking adult males.

Growing Old

We've already seen how the reproductive lives of wild chimpanzees play out, with females displaying a reproductive life span markedly different from that of humans. A female chimp's reproductive system ages at about the same rate as the rest of her body. This is in stark contrast to a human female, whose reproductive system ages and fertility ceases in midlife but whose body ages only decades later. But aging in wild chimpanzees is still poorly understood. Gerontological studies that might provide important comparative information for human aging are mostly limited to captive chimps. Most chimpanzees that die in field studies simply disappear one day and are never seen again. The cause of death is rarely known, except in cases in which the chimp was badly injured or visibly very sick. Captive chimpanzees spend their lives eating nutritious food, for which they never have to forage. Their diet and inactivity make couch potatoes out of even the healthiest zoo chimpanzees. They spend their lives under the care of dedicated keepers, nutritionists, veterinarians, and behavioral therapists. They have no predators to worry about and no epidemics to succumb to. They grow at a rapid rate to a large size compared to wild chimpanzees and live much longer on average.

In the wild, chimpanzees have a life expectancy at birth of only fifteen to nineteen years, based on studies of mortality rates at Kanyawara, Gombe, Taï, Mahale, and Bossou. But if a wild chimp survives to sexual maturity, he or she can expect to live another fifteen to twenty-four years.[24] These figures are substantially lower than survivorship of the average zoo population. Because field studies began only in 1960, there is no way to precisely age any wild chimpanzee past its midfifties, and for most shorter studies the maximum life span in the wild is unknown. But we have never had an authenticated case of any wild chimpanzee reaching sixty years of age (perhaps partly because field studies only

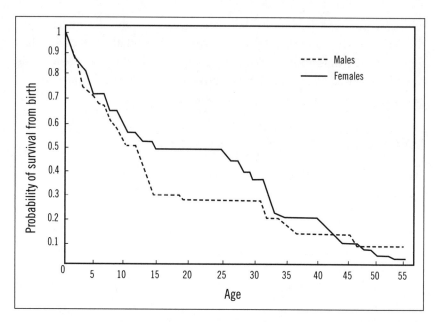

Comparison of age-specific survival probability from birth, females versus males, Kanyawara.

began fifty-seven years ago). In captivity, chimpanzees have been documented to live to their midsixties (sixty-six is the oldest known age currently), and females may reproduce until not long before that age.

Life expectancy and mortality data take decades to compile before meaningful conclusions can be drawn. We have information for many human populations on how long people live and what risk factors affect their life span. Such information has until recently not been available for chimpanzees because they too live long lives, and compiling information in the wild is difficult. Recently, Martin Muller and Wrangham examined the population biology of the Kanyawara chimpanzees. They found that the mean overall mortality was just under 4 percent annually, much lower than the estimates for other wild chimpanzee populations but twice the figure for human hunter-gatherers. Life expectancy at birth was four and a half years higher for females than for males (twenty-one and a half versus seventeen).[25] Chimpanzee populations show wide variation. At Ngogo, only ten kilometers away from Kanyawara, Brian Wood of Yale University and colleagues found mortality rates that were much lower and similar to what we see among zoo chimpanzees. They attribute this to the extremely productive forest at Ngogo, which may lower death rates attributable to undernourishment and the diseases and stress that are normal at other sites.[26]

Disease outbreaks happen regularly enough among wild chimpanzees to strongly influence life expectancy. Illness is the major cause of death among wild chimpanzees, when cause of death can be ascertained. Other sources of mortality—accidents, predators, and fights within or between communities—vary widely from one population to the next and from one decade to the next. Outbreaks of respiratory infections, debilitating internal parasites, polio, mange, and simian immunodeficiency virus (SIV) have all hit the Gombe chimpanzees over the past five decades. In a series of papers, Thomas Gillespie of Emory University, Karen Terio of the University of Illinois, Brookfield, and colleagues documented a variety of nematode and other intestinal infections in the Gombe chimpanzees, some of which were apparently harmless and others fatal.[27] In Mahale, a disease with flu-like symptoms swept through M group in 2006, killing as many as twelve chimps, nearly one-fifth of the community. Nine more disappeared and may have been victims of the outbreak as well. In all, a third of the community members showed symptoms of being sick. The outbreak resembled a fatal epidemic in 1993 that had killed at least eleven members of M group. The Mahale researchers believed the outbreak spread into M group from the local human population: scientists, tourists, or the national park staff.[28]

In West Africa, the Taï chimpanzees have experienced population crashes due to outbreaks of anthrax—possibly spread from livestock outside the forest—and the Ebola virus. Ebola has in all likelihood more severely affected chimpanzees and gorillas across western and central Africa than it has any human population. The source of the disease is thought to be fruit bats, but as of 2017 that is still unconfirmed. The main study community at Taï suffered a long-term decline from eighty to twenty animals during the 1990s and early 2000s as a result of killings by poachers and local farmers.[29]

We're still learning about the scope, history, and impact of SIV. While SIV does not inevitably progress to a fatal syndrome of illnesses comparable to AIDS, it sometimes eventually results in AIDS-like symptoms, and it does increase mortality in wild chimpanzee populations. There are

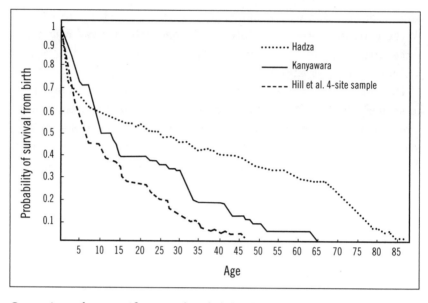

Comparison of age-specific survival probability from birth, chimpanzees versus hunter-gatherers.

other emerging viruses that have yet to be identified that contribute to chimpanzee mortality. As long as chimps continue to exist, additional pathogens and health issues will no doubt continue to emerge. Muller and Wrangham believe that most of the mortality risks that wild chimpanzees face today are from novel sources. It's interesting that, although we have seen epidemics cause some chimpanzee communities to plummet in size, these drops have been relatively short term.[30] Nearly all the communities that we've studied for decades have remained more or less stable in size. The Kahama community at Gombe and K group at Mahale are the best-known exceptions; both were extirpated by conflict with neighboring communities. No communities have been destroyed by any extrinsic cause such as disease outbreaks. Increasing human populations surrounding chimpanzees' forest homes heighten the risk of disease transmission from local people to the apes. But more people in the area means fewer predatory species that kill chimps. At many long-term chimpanzee studies, death by lion or leopard is a thing of the past. Recent human contact with chimpanzee populations on a large scale has without question been bad for chimpanzees overall, but once human factors are introduced, the unintended consequences don't always follow directly from the altered conditions.

Wild chimpanzees rarely die of old age. As aging takes its toll on health and vigor, a leopard, an infection, or a disease outbreak finishes off the ape's life. But survival of a wild chimpanzee to old age follows a different pattern from that of hunter-gatherers. Humans live three decades longer. Human females live those three decades in a postreproductive state. Some of the most prevalent human diseases that occur later in life, such as most cancers, are barely known in chimpanzees. Like women, female chimpanzees live longer than males, with life expectancy differences between the sexes beginning at puberty. At every life stage, chimpanzees suffer higher mortality rates than human populations that have been studied. As much as we look to chimpanzees for models to explain humans, we have to be cognizant of the evolutionary gulf between us.

CHAPTER 7

Why Chimpanzees Hunt

THE HUNT is often heard before it's seen.

You're walking single file along a forest path behind a party of male chimpanzees. It feels like tagging along on a Boy Scout hike. The chimpanzees abruptly freeze. They have a habit of stopping in midstep, one leg suspended in the air behind them. They stand frozen, staring ahead on the trail and up into the trees. You don't see or hear anything, but they clearly do. Then you hear leaves and twigs breaking overhead some fifty yards up the trail. There are anxious, high-pitched chirps, the alarm calls of red colobus monkeys. They've seen or heard the chimpanzees coming and are closing ranks in a panic. The male chimpanzees bolt up the trail, their eyes glued to the treetops. You follow them and arrive at a chaotic scene. There are male chimpanzees in tree attacking the monkeys. Other chimps are on the forest floor, gazing at the action overhead. There are usually one or two males who are catalysts for the hunt. Their eagerness to eat meat outweighs any fear of being bitten by their prey, and others will follow their lead.

The monkeys are under siege. Mothers have gathered up their babies and the adult males are trying to position themselves between the chimpanzees and the female monkeys. But the number of attackers and the tactics they employ mean that one or more monkeys will likely be caught. The male colobus fight the chimpanzees fiercely and are often successful, especially when the chimps are few in number. They leap onto the male chimpanzees, biting them. Some of the chimpanzees flinch when counterattacked; others are undaunted and charge through the melee to grab their prey. Sometimes the chimpanzees capture more than one monkey in a hunt, occasionally many.

The hunt is often so chaotic that you aren't sure of the outcome until a male chimp emerges from a thicket carrying a dead monkey in his hand or mouth. Hunts can be gruesome to watch, and, if you're collecting data for research, you experience the fascination of the hunt while dreading the awful outcome for the monkeys. I once watched Prof, an older adult male, pursue a female colobus from treetop to treetop. When she finally gave up her flight, her body was quaking. Prof reached her, then extended his hand to her belly, where a baby colobus was clinging to its terrified mother's torso. The mother, too exhausted to fight or flee, seemed to surrender her baby to the hunter. Prof took the tiny creature and killed it with a bite to the skull, then departed, leaving the mother monkey to live another day.

The hunt often ends as abruptly as it started, and the chimpanzees sit down to eat their catch. Intense politicking begins. One or a few big males typically control the meat. Sometimes the males try their best to avoid sharing scraps with anyone. At other times, there is a barter system at work. Males share willingly with their mothers and brothers, less so with nonrelatives. They openly snub rival males, sometimes leading to charging displays by visibly frustrated males who are left out of the prize sharing. Eventually, everyone settles down and the ensuing hour or more is spent contentedly crunching bones, sucking marrow, and chewing on colobus meat.

This same scenario plays out in many forests across Africa. The details vary since hunting strategies vary among communities. Hunting is one of the cultural traditions intrinsic to chimpanzee life. Hunting and meat sharing thus become another way in which we see the roots of human behavior when we watch chimpanzees.

One morning in late 1960, not many months after she first arrived at Gombe, Jane Goodall was sitting at her favored observation point, a hilltop knoll she called the Peak, when she heard a commotion in the valley below. She spotted a chimpanzee holding something in his hands. It was a baby bushpig, a boar-like wild pig common in East Africa. The chimpanzee had apparently snatched one from its mother and was now munching on a pork breakfast. Along with the meat, the killer was plucking and eating leaves. Goodall's observation was the first evidence that our closest kin consumes the meat of other mammals.[1]

In the early 1960s, when chimpanzees were assumed to be vegan, this was a heretical thing to say. Skeptics first claimed that Goodall was simply mistaken. Later, when it became clear that hunting and meat eating are routine behaviors, critics charged that Goodall's proximity had somehow tampered with natural chimpanzee behavior. Not until hunting was seen in numerous other forests did the entire scientific community accept that meat eating is a core aspect of chimpanzee behavior.

The Hunting Ape

Despite the avidness and frequency with which chimpanzees hunt, they are by no means carnivores. Neither are they the herbivores they were once believed to be. They are omnivores, like us, although they consume a smaller amount of meat than do most human populations. The exact quantity of meat consumed is difficult to measure; you can't ask a chimpanzee for permission to weigh his meal before he eats it. But meat composes only 1 to 3 percent of the diet, based on the amount of time they spend eating.

Chimpanzees are the only great ape that is a systematic predator (though some would argue that we've underappreciated bonobos in that role). The list of prey eaten by chimpanzees is diverse and prominently features other primates. This is likely because monkeys are the most available, catchable mammalian prey that are active in the daytime in many forests. Any animal that crosses a chimp's path is a potential meal. Not all potential prey species are eaten, however. I once saw a foraging party of chimpanzees discover a litter of baby civets—small, raccoon-like carnivores—in a thicket. Each of several male chimps ended up with a civet carcass, which they carried around all day long, flung over their shoulders or dragged in one hand. One male even took the carcass to bed with him, discarding it in his nest when he awoke the following morning. The chimpanzees didn't regard the civets as food, although they're perfectly edible; I've shared a meal of civet meat with a band of human hunter-gatherers in East Africa. Civets possess musk glands—from which some perfumes are made—so perhaps the chimpanzees don't care for the taste of their meat.

Although Goodall observed many chimpanzee hunts during the early and mid-1960s, the first field study that focused explicitly on meat eating

Species of mammalian prey eaten by wild chimpanzees across equatorial Africa. If the list were expanded to include all forms of vertebrate prey, it would include at least forty species.

Common Name	Scientific Name
Primates	
Red colobus monkey	*Procolobus badius*
Black-and-white colobus	*Colobus polykomos*
Savannah baboon	*Papio anubis*
Blue monkey	*Cercopithecus mitis*
Red-tailed monkey	*C. ascanius*
Diana monkey	*C. diana*
L'Hoest's monkey	*C. l'hoesti*
Gray-cheeked mangabey	*Lophocebus albigena*
Sooty mangabey	*Cercocebus atys*
Potto	*Perodicticus potto*
Senegal galago	*Galago senegalensis*
Vervet monkey	*Chlorocebus aethiops*
Patas monkey	*Erythrocebus patas*
Nonprimates	
Blue duiker	*Cephalophus monticola*
Bushbuck	*Tragelaphus scriptus*
Bushpig	*Potamochoerus larvatus*
Banded mongoose	*Mungos mungo*

* Data sources: Stanford 1998; Wrangham and van Zinnicq Bergmann-Riss 1990

was carried out by Pennsylvania State University graduate student Geza Teleki. He spent a year in Gombe in the late 1960s and witnessed thirty predatory attempts by chimpanzees. A total of four colobus monkeys and a dozen baby baboons were killed. His descriptions of meat eating and meat sharing were the best evidence at the time about this newly discovered behavior.[2] Teleki's results were, in hindsight, however, not representative of chimpanzee meat-eating behavior. In the 1960s, Goodall cleared an area of forest and began putting out bananas to draw the apes to a place where they could be observed and filmed up close.[3] The clearing soon became a focal point of activity.

In all likelihood, the forest clearing temporarily created a new favorite prey. Chimpanzees encounter baboons in the forest more often than any other mammal, and they have likely always hunted baby baboons. The problem in catching them is that they're often protected by big male baboons, which possess formidable canines. Baboons were drawn to the new chimpanzee feeding station too, and suddenly there were many more baboons and their infants in the immediate vicinity of the chimpanzees. During the first decade of Goodall's work at Gombe, nearly 40 percent of all predations observed were of baboons, peaking in 1969.[4] Once researchers began to spend more time in the forest instead of the feeding station, observations of predation shifted mostly to red colobus, and they remain so today. When Curt Busse studied chimpanzee hunting patterns only a few years after Teleki's work, nearly all of his observations were of red colobus hunts in the forest.[5]

The Teleki study nevertheless provided the most detailed evidence at the time about chimpanzee meat eating. He noted the relationship between the size of the chimpanzee hunting party and the likelihood of hunting success. He provided detailed information on the sequence of carcass consumption—which bits were eaten in which order—which is of great interest to archaeologists studying early human meat eating. Teleki also documented meat eating after the kill, which is for many scientists the most fascinating and revealing aspect of the hunt. It was Teleki who first suggested that the sharing of meat might be motivated by social factors such as nepotism and kinship. He saw chimpanzee meat eating as an intensely political behavior, possibly involving cooperation among males and nepotistic motivations for meat sharing afterward.[6]

Teleki's small sample of hunting data, combined with the observations on predation that Goodall herself had already made during the 1960s, laid the groundwork for all studies of chimpanzee hunting that followed. His early descriptions of hunts and their aftermath were in my mind as I watched my first hunts at Gombe twenty years later. He understood that, although the primary motivation for hunting may be calories and nutrients, chimpanzees' intense interest in meat goes beyond nutrition; meat is also a useful political tool. As Goodall noted before him, Teleki saw chimpanzee hunting and meat eating as a proxy for the same behaviors in early hominins. It is the closest we will ever come to a glimpse of our earliest apelike ancestors as they evolved from herbivores to omnivores.

A few years later, what Toshisada Nishida and his colleagues learned about hunting in the Mahale Mountains was a revelation. Skeptics of Goodall's work believed that meat eating might be an anomalous behavior caused by a researcher's presence. Nishida's observations made it clear that hunting is a routine and normal part of chimpanzee biology.[7]

Through the first decade and a half of chimpanzee field research, all the observations had been made in East Africa, in forests that were seasonally very dry. In the late 1970s, Swiss-born primatologist Christophe Boesch established his chimpanzee study in Taï National Park, Ivory Coast. In a rain forest habitat very different from the drier sites of Gombe, Mahale, and Budongo, Boesch was rewarded with a previously undescribed pattern of hunting. Taï chimpanzees cooperated during hunts to a degree not seen in the sites in East Africa. They also preferred adult monkeys as prey instead of juveniles, in stark contrast to the chimpanzees at Gombe. Females played a more important role in the hunt than observed elsewhere.[8]

It was at this juncture that I began a career in great ape research, having been invited by Goodall to conduct a postdoctoral study at Gombe. I focused on the predator-prey relationship between the chimpanzees and red colobus. By following two colobus groups and learning to identify individuals in them, I was able to watch more than one hundred encounters between the chimps and colobus, which resulted in sixty hunts. At the time, this was the most systematic analysis of the interaction between nonhuman primate predators and prey. My work showed the profound impact that chimpanzee predation could have on prey populations. It also indicated a strong social role for hunting in the nepotistic distribution of meat after a kill. When my first results were published during the 1990s, the debate that ensued tended to overshadow the overall research, which was about how chimpanzees behave as predators and how the colobus respond to the threat of predation.

In nearly every forest in Africa in which chimpanzee hunting behavior has been studied, the red colobus monkey is the most frequent prey (Budongo Forest Reserve in western Uganda, which lacks red colobus, is the exception). At Gombe, adult male red colobus are very rarely taken, but far to the west at Taï they are the most frequent prey. An adult male colobus might weigh nine kilograms. A large bushbuck fawn, weighing in at around twelve kilograms, is among the largest prey size recorded.

Field studies focused specifically on chimpanzee hunting behavior

Researcher(s)	Dates	Site	Key Publication(s)
Geza Teleki	1968–1969	Gombe, Tanzania	Teleki 1973
Curt Busse	1973–1974	Gombe, Tanzania	Busse 1977
Christophe Boesch*	1979–	Taï, Ivory Coast	Boesch and Boesch 1989
Craig Stanford	1991–1995	Gombe, Tanzania	Stanford 1998; Stanford et al. 1994a, 1994b
David Watts/John Mitani*	1993–	Ngogo, Tanzania	Watts and Mitani 2002
Ian Gilby/Richard Wrangham	1990–2003	Kanyawara, Uganda	Gilby and Wrangham 2007
Ian Gilby	1999–2002	Gombe, Tanzania	Gilby 2006; Gilby et al. 2010

* Studies included data from a larger study of chimpanzee behavior.

In some forests, male chimpanzees mainly eat baby colobus the size of kittens. I have seen Gombe chimpanzees capture and kill adult males and then discard them to pursue other prey. The colobus carcass falls to the forest floor and becomes a meal for a chimpanzee who had been a spectator at the hunt up to that moment. Why would a hunter pass up the meat of an adult in favor of the tiny quantity of meat provided by an infant? We can't be sure, but it suggests that hunting is motivated by more than a desire for nutritional and caloric content. The tender baby may simply taste better to the chimpanzee's palate or be more easily eaten than the adult. In Taï National Park, however, adults are the favored prey, an example of the fascinating variation in hunting strategies that exists among study sites.[9]

Once a monkey has been captured, the chimpanzee must dispatch it, or at least he must immobilize it before beginning to eat it. Sometimes a chimp begins to tear the monkey limb from limb while it is still struggling to escape his grasp. When a chimpanzee catches an adult colobus, he typically thrashes its body against a tree limb or the ground, killing or mortally wounding it. Baby monkeys are killed by a quick bite to the skull or neck. Although chimpanzees in many sites prey mainly on immature monkeys weighing no more than a kilogram, the total biomass consumed can be substantial because of the number of colobus that are killed. Based on my estimates of colobus body weights, the Kasekela chimpanzee community consumed a total of more than five hundred kilograms of colobus monkey in some years. Adding in the other prey species also consumed yields an estimate of seven hundred kilograms of vertebrate prey.[10] Although this sounds like an enormous amount of meat eaten by an ape that is supposed to be no more than occasionally omnivorous, the estimate is in line with earlier calculations made by Richard Wrangham and Emily van Zinnicq Bergmann-Riss. They estimated, also at Gombe, that the Kasekela community consumed 441 kilograms annually during the 1970s.[11]

I was especially interested in how much meat was eaten by each individual per day, because I wished to gauge exactly how important meat was in the chimpanzee diet. This sort of estimate had not been made for chimpanzees, partly because it's difficult to estimate visually how much meat they eat. It would also shed light on the nature and pattern of meat eating in apelike fossil humans, about which we know little. At Gombe, during the peak hunting months of August and September, the mean daily intake of meat—really carcass biomass, since all parts of the carcass are eaten—by adult and adolescent chimpanzees ranged from thirty-two to ninety-seven grams in different years, with a mean of sixty-five grams consumed daily. This may not sound like a lot of meat—a fast-food burger contains much more—but considering that at the time we didn't think that chimpanzees ate much meat at all, it was a revelation. Some hunter-gatherer societies eat little more than that amount of meat during their leanest months.[12]

As I noted earlier, hunting shows a seasonal pattern at Gombe and some of the other chimpanzee study sites, although the hunting peaks tend to be modest and not necessarily statistically significant. In the Mahale Mountains, hunting is sharply seasonal; in Gombe and Taï less so. Both hunting frequency and hunting success may vary from month to month. The average hunting success rate in the 1990s varied from 36 percent to 84 percent. This fluctuation didn't appear to be due to any social parameter such as party size or the number of males in the hunting party.[13]

We use chimpanzees to understand the origin of meat eating among early humans, but there are important differences that limit the comparison. Chimpanzees have never been observed killing prey animals larger than an adult monkey or young antelope. Even when confronted with large carcasses of recently dead animals such as bushbuck antelope, which would provide them with more meat than they could catch in a month of hunting, they tend not to treat the carcass as a potential food source. Chimpanzees have never been observed trying to bring down large prey with weapons of any kind; they simply don't have the means to kill a larger animal that has horns or tusks to defend itself. Only very rarely in the history of chimpanzee field research have the apes been observed to use tools to assist in capturing even small prey. Since 2007, Jill Pruetz and her colleagues have been describing a tool-assisted pattern of meat procuring among chimpanzees. At Fongoli in southeastern Senegal, her chimpanzees use sticks as bludgeons to assist in injuring and extracting galagos (bush babies) from the tree cavities where they sleep during the day. Such meat eating might better be called extractive foraging than hunting. Notably, females account for a larger percentage of hunts at Fongoli than they do at any other study site. It's one of the more intriguing observations of chimpanzee behavior in recent years.[14]

Only when chimpanzees hunt monkeys do they appear to engage in planning and tactical decision making during the chase. By studying the ways in which chimpanzees use meat to manipulate the behavior of others, we may see the roots of a humanlike intelligence not visible in social carnivores such as lions and wolves.

Although lions and wolves are armed with far more formidable weapons than chimpanzees are, chimpanzees are more efficient predators. In some forests more than 80 percent of hunts are successful, a figure not matched by any of the social carnivores except African wild

Predation efficiency of chimpanzees compared to other social predatory species.

Species	Percent Hunts Successful	Source(s)
Chimpanzee	54–82	Stanford 1998; Watts and Mitani 2002
Gray wolf	15–64	Mech et al. 2001; Sand et al. 2006
Lion	14–46	Funston, Mills, and Biggs 2001
Cheetah	11–23	Mills, Broomhall, and du Toit 2004
Spotted hyena	33	Holekamp et al. 1997
African wild dog	80	Fanshawe and Fitzgibbon 1993

dogs. This success is related to the group nature of chimpanzee hunting. The more hunters there are in the hunting party, the higher the odds of a successful hunt. Most large social predators such as lions and wolves are carnivores. Their daily decision is where to find meat and how to catch it. But chimpanzees are omnivores, and although they relish meat, it's only a tiny percentage of their diet. They decide on a daily basis whether and when to hunt instead of continuing to search for fruit.

Hunting is a seasonal behavior for many chimpanzee populations. This is fascinating because of evidence that early humans also ate more meat at some times of year than others. In Olduvai Gorge in Tanzania, the dry season appears to have been a time of intensive meat eating by early hominins three million years ago.[15] The same is true at Gombe, which is only a few hundred kilometers away from Olduvai. At Gombe, hunting peaks during the long dry season from July to October, although our perception may be skewed because the chimpanzees are more easily followed and observed in the dry season, when there are fewer leaves on the trees. Research on hunting at Gombe showed that hunting is least frequent during the height of the rains in April and May.[16]

In addition to monthly variation in hunting frequency, chimpanzees are prone to bursts of intensive hunting. Goodall noticed this in the 1960s—she labeled them "hunting crazes"—and was unsure what caused them. After a period of infrequent hunting that might last weeks, the chimpanzees would abruptly begin to hunt every colobus group they encountered, sometimes killing monkeys on a daily basis. These episodes of intense hunting are no doubt influenced by how often the chimpanzees encounter colobus monkeys. I found that the Gombe chimpanzees decided to hunt about two-thirds of the time that they met colobus in the forest.[17] But at Ngogo in Kibale National Park, Uganda, only about one-third of encounters with colobus result in hunts.[18] Why not hunt colobus every time they're encountered? The decision to hunt, at least at Gombe and Kibale, appears to be strongly influenced by the number of male chimpanzees in the foraging party and the identity of the hunters. Some males are avid and skilled hunters who serve as catalysts for hunting, while others are less eager and more easily intimidated by the male colobus. My study at Gombe and a later analysis by Ian Gilby of Arizona State University and his colleagues at Kanyawara found that when the catalyst dies, the frequency of hunting may drop.[19]

Despite having only their hands as weapons, the power of working together sometimes enables a hunting party to kill many monkeys in the same hunt. Most successful hunts of red colobus yield one kill, but at times two or more are taken. During my own study, seven monkeys were once killed in the same hunt, and up to a dozen kills have been observed in other forests. Since most red colobus groups in the core hunting area had between twenty-five and forty animals, the chimpanzees were capable of decimating a group in a single hunt. Research at Ngogo has shown much the same effect. David Watts and John Mitani suggested that chimpanzees may exert enough hunting pressure to bring about a long-term colobus monkey population decline.[20] Chimpanzees, despite including only a small quantity of meat in their diet, can exert a profound impact on their favorite prey.

Sex differences in hunting behavior are stark; males tend to hunt together, and females are often carrying young infants that are vulnerable to colobus bites. When I was preparing to set off to study hunting in the early 1990s, Goodall advised me to pay close attention to female chimpanzees; she felt their role in hunting had been underappreciated. I went to Gombe expecting to see females more involved in the hunt than had previously been reported. It's not easy to follow female chimpanzees. Following male chimpanzees can be tough work, as they travel rapidly in rugged terrain. But due to their quieter and more solitary ways, lone females may slip away from a researcher, after which they're more difficult to find than noisy, garrulous parties of males.

It turned out that there was actually less hunting by females than I had expected. About 90 percent of kills were made by males during my Gombe study period. This figure varies from one forest to the next—at Mahale it is about 80 percent, at Taï about 70 percent—but it's clear that hunting is a largely male activity.[21] At Ngogo, adult and adolescent males account for fully 98 percent of all kills.[22] In fact, the difference in the importance of meat in the diet of males versus females is striking enough that Geraldine Fahy of the University of Kent and colleagues found meat eating left a biochemical signal in hair and bones of male Taï chimpanzees but not in females.[23] Males do not only value meat as a nutritional supplement; they also use it as a political bargaining chip. Females are equally eager to eat meat, perhaps even more so given that they bear the metabolic costs of reproduction. But female chimpanzees do not cooperate well with one another. So when hunting, they are usually alone or with one or two others, which does not bode well for a successful hunt.

After a kill, females often beg for meat scraps from males. If they are high-ranking females allied with or related to the male who possesses the meat, they will likely receive some scraps. If they are younger adult females without offspring—that is, potential mates—they will also likely receive meat. It is typically up to the male who controls the carcass to decide whether a female will be allowed to consume some. This does not mean that females do not wield power and influence surrounding meat eating. Female reproductive cycles may exert a strong indirect influence over male hunting behavior.

Whether chimpanzee foraging parties wake up in the morning with the intention of finding prey has been a subject of much debate. Most chimp researchers have concluded that hunting parties more or less randomly encounter prey in the course of foraging for fruit. This is unusual behavior for social carnivores, which wake up each morning wanting to find meat and little else. Most traditional human hunters also have a plan for the day that involves finding meat, and they gather plant foods and other edibles as they search. I saw this myself in a day spent with Hadza foragers in Tanzania. From dawn to late afternoon, a hunting party of four men walked or jogged across the scrubby plains. They collected bird eggs and baby birds and stuffed them unceremoniously into their pockets. They also stopped to pick occasional plants, but their main quarry of the day was meat. When they finally located and killed a civet, they regarded the day as productive. Chimpanzees do essentially the opposite. In a day mainly spent finding and eating fruits and leaves, they will sometimes hunt monkeys and other mammals they encounter along the way.

There are a few researchers who feel they have evidence that chimpanzees search for meat, beginning their day with the intent of finding red colobus monkeys. Boesch has claimed that Taï chimpanzees make detours in the forest when they hear distant monkey calls; he believes that many colobus encounters are intentional, not accidental. But it's more likely that the chimpanzees are attracted to forest areas that colobus frequent, which may represent an indirect intent to find them. Even if the chimps are not searching for monkeys, they are likely aware of parts of their range in which prey are often encountered. Perhaps individual chimpanzees have memories—pleasurable ones, no doubt—of successful hunts when they enter the areas where those hunts took place.

Scavenging, the consumption of animals that are already dead, is an occasional source of food. Chimpanzees scavenge, but they also often ignore carcasses they encounter. Watts estimated that the Ngogo chimpanzees encountered a carcass once every three to four months but were only observed to eat scavenged meat four times in eleven years.[24] I once carried an uneaten red colobus carcass back to camp after a hunting party killed and ate four other colobus from the same group. I placed the carcass in the banana feeding station clearing and waited for a chimpanzee to walk by. I predicted that a female chimpanzee would be more likely to scavenge than a male, because males are more easily able to hunt for meat. The miniexperiment lasted an hour. The first chimpanzee that arrived was a mother with her infant; she whisked the carcass up a tree to eat it and shared the meat with her baby.

Other instances demonstrate that scavenging doesn't necessarily happen even when it seems to make eminent sense. In one instance in Gombe in 1988, a foraging party of chimpanzees came across a freshly killed bushbuck carcass. The antelope's abdominal cavity had been emptied of its viscera not long before; it was likely that a leopard had made the kill. All the chimpanzees were extremely curious, and they nibbled small scraps of meat from the carcass. But despite the bounty of protein and fat that this forty-kilogram animal represented, the chimpanzees mainly toyed with it. At one point a chimp actually crawled partly inside the hollowed-out ribcage and rolled playfully around in the dead antelope's body cavity. After two hours the party had lost interest and moved on. None of the chimpanzees had consumed more than a token amount of meat from the dead antelope.[25]

Scavenging has been reported at all the major chimpanzee study sites, but it is nowhere common. Perhaps chimps have a tendency to avoid dead sources of meat to avoid the unintentional ingestion of decomposing food and bacteria. Or perhaps it just highlights that meat eating has much to do with the social and political aspects of the hunt and not simply the consumption of nutrients and calories. If meat eating were only about nourishment, then scavengeable meat should be a prime menu item, as it is for lions, wolves, hyenas, and most other large carnivores.

Why Do Chimpanzees Hunt?

Explaining why chimpanzees hunt would seem straightforward. Meat provides calories, protein, animal fat, and essential amino acids. But it's not that simple. Estimating meat's caloric importance is difficult, but as a percentage of the time they spend eating overall, it might be 1 to 3 percent. There is at least one long-term chimpanzee study—Budongo—in which chimpanzees consume practically no meat at all, with no evidence that they suffer nutritionally.[26]

A few researchers dispute the idea that chimpanzee hunting is motivated by factors other than calories and nutrients, but most agree that meat plays a role in chimpanzee society that goes beyond nutrition. They base this conclusion on aspects of the hunt and the aftermath that appear to refute a purely energetic cause. Recent studies have considered the value of the meat of mammals relative to other sources of animal nutrition. Claudio Tennie of the University of Birmingham in England and his colleagues analyzed the nutrient value of scraps of mammal meat versus insect protein, since male chimpanzees invest a great deal of time and energy in acquiring both in some seasons. They concluded that, although we don't know which nutrients chimpanzees seek when they pursue meat, it's likely that the two kinds of prey consumption complement each other. Hunting other mammals yields the essential amino acids that may compensate for the considerable time invested and the small quantity of the reward received.[27]

Of course, even if meat is valued by a chimpanzee for something other than its nutritional qualities, the fundamental importance of meat stems from those qualities, which are what makes it a valuable tool for social barter. But what exactly do chimpanzees gain from meat? Let's consider the major explanations.

Calories. A chimpanzee's daily life is dictated by the search for ripe fruit. Finding ripe fruit requires walking kilometers each day, and the calories burned must be balanced by calories consumed. The nutritional and caloric values of chimpanzee foods have been estimated for some populations but not all, and in general we know little about how the animals avoid energy deficits. Meat is not an ideal source of calories, because finding it is unpredictable and not a daily occurrence. Might chimpanzees therefore hunt to supplement the calories they harvest from plants?

This was first suggested by the researchers in Mahale Mountains National Park, who hypothesized that chimpanzees hunted more in the dry season because of shortages of other foods.[28] The problem with this idea is that there was no real evidence from Mahale that food was actually less available in the dry season. While many trees shed their leaves at that time, some are laden with fruit. As we lack quantitative studies of food availability and its caloric content, any connection between hunting and lack of other foods is speculation.

A more concrete connection was made by Martin Muller while studying chimpanzees at Kanyawara. He used an indirect measure of food shortage and nutritional stress by studying ketones in wild chimpanzee urine. When stored body fat is depleted, fat molecules are broken into fatty acids. The body uses these fatty acids in place of glucose, and ketones are formed. When food was less available, Muller hypothesized, chimpanzees would be calorically stressed and should begin to burn their own fat, a process called ketosis. The evidence of this energy stress would show up in the form of ketones in chimpanzee urine samples. Muller showed that in some seasons, the chimpanzees clearly did not have enough to eat, based on their urinary ketone levels. There was no link, however, between their seasonal energy shortfall and the frequency of hunting.[29]

Unfortunately for the idea that hunting may occur to offset plant food shortages, some researchers have argued that chimpanzees actually hunt *more frequently* when there is more plant food available. Watts and Mitani found that Ngogo chimpanzees ate more meat when fruit was abundant.[30] Ian Gilby and Wrangham, working in nearby Kanyawara, also found that hunting was correlated with periods of fruit abundance rather than with fruit scarcity, and they suggested that Kanyawara chimpanzees hunt when they can afford to risk wasting the energy on failed hunts because they have plentiful other foods available.[31] At Ngogo, times of great fruit availability were also times of large chimpanzee party size; I had published results in several paper in the 1990s showing that large parties typically hunt more frequently and more successfully.

Protein. It could be that chimpanzees hunt in hopes of acquiring a particular key nutrient rather than just energy (in the form of calories) in general. If that were the case, protein would be a prime suspect. It's a macronutrient high on the wish list of nearly every animal, whether carnivore, omnivore, or herbivore. Meat is of course protein rich, but protein is also available in a tropical forest in plant foods. Wrangham, Nancy Conklin-Brittain, and their colleagues showed that the diet of Kanyawara chimpanzees was high in protein, obtained from both fruits and leaves in addition to the small fraction of meat in their diet.[32] This wealth of protein was also found in the diets of females and juveniles, who consume little meat. Protein might therefore be a less likely driving force behind hunting, at least at Kanyawara.

Fat. Unlike protein, fat is not easily obtained in a tropical forest. Saturated fat in particular occurs in very few plant foods. But it's readily available in the carcass of any mammal, along with all the essential amino acids. There is an easy way to test whether fat might be a primary force driving hunting. When a chimpanzee makes a kill, does he consume the fat-rich parts of the monkey's body first or not? The brain and bone marrow are the two most fat-dense parts of the body; where do they rank in the sequence of carcass consumption? The answer is, it depends on which forest you're in. In Gombe, a chimpanzee that makes a kill almost always eats the brain first. He will crunch through the skull of a small monkey or suck the brain matter out of the skull in a larger animal. In West Africa, Taï chimpanzees prefer bone marrow over brain tissue, and they extract it before eating any other part of the carcass. In both sites, fatty parts of the carcass are preferred.[33] So, solely based on Gombe and Taï, fat does seem to be a highly sought nutrient in the prey. Unfortunately for the hypothesis, the brain and long bone marrow are not necessarily eaten first in all other sites.

A further intriguing aspect of fat consumption and hunting is that saturated fat is available in one very abundant plant food at Gombe: the oil palm. Oil palm trees were introduced from West Africa long ago and are planted in groves so that their fruit can be harvested and turned into cooking oil. They occur abundantly in parts of Gombe, dating from the days when the forest was dotted with villages. Palm fruits are among the most important foods in the Gombe chimpanzee diet, and they are eaten in every month of the year. They are actually fattier (in fat grams per unit of weight) than monkey meat, which tends to be lean. Unlike colobus monkeys, palm trees are easy to locate, they don't flee from the chimpanzees, and they don't bite back when attacked. So why would a chimpanzee engage in high-risk, unpredictable hunts when he could sit in a palm all day long, all year round, eating fatty palm fruits? This is one piece of evidence that suggests that hunting is not driven purely by nutritional factors.

Trace minerals and elements. It's impossible to test the importance of rare components in the chimpanzee diet, but some are intriguing to speculate about. I was once approached at a conference by a spry, elderly Australian man who came up to me and simply said, "It's salt." I was reminded of the opening scene in the film *The Graduate* when a party guest tells young Dustin Hoffman to think about "plastics." The Australian man, it turned out, was a world-renowned authority on the physiology of salt in the human diet. A carcass is an excellent source of salt, which is difficult to obtain elsewhere in a largely vegan diet, he argued, so I should consider salt as the driving force behind the chimpanzees' desire for meat. I wasn't sure how to respond, except to ask how I should try to isolate this one item in the diet from all the others in a wild animal that can't be subjected to laboratory experiments.

Likewise, it has been suggested to me that chimpanzees might eat colobus monkeys to provide themselves with a B-vitamin boost. Some members of the B-vitamin complex cannot be derived from plant foods. They can, however, be obtained from eating the digestive system contents of certain herbivorous mammals. Colobus monkeys, with their unusual ruminant-like digestive systems, are among these B-vitamin-rich mammals. It seems unlikely to me that chimpanzees would go to the considerable time and effort, not to mention risk of injury, for the sake of a vitamin. But this hypothesis is easily tested. Do chimpanzees prioritize eating the stomach and intestines of colobus when they capture them? In fact, the digestive tract and its contents are usually among the last items consumed after a kill. I've watched a chimpanzee squeezing the contents of the intestines out like toothpaste from a tube while consuming it. But it's done after all the preferred body parts are already eaten. So there's little evidence from the sequence of carcass consumption to suggest an important role for B vitamins as a driving force behind chimpanzees' hunting.

Other trace substances are intriguing but as highly speculative as salt. Copper, zinc, and other trace minerals could be sought in a prey animal's carcass, but would they be the primary reason the apes hunt? It seems unlikely. This idea gained some traction in the work by Tennie and colleagues described earlier. They argued that instead of maximizing caloric or macronutrient intake in hunting, chimpanzees maximize their odds of acquiring even small scraps of meat in order to consume micronutrients in them.[34] If cooperative or group hunting increases the odds that an individual will obtain even a tiny quantity of meat, it might explain why chimpanzees invest hours in hunting or begging for meat when the return is often seemingly so meager.

Could Hunting Be about More than Nutrition?

Since the first observations of chimpanzee hunting, many primatologists have believed that the pursuit of meat has a social function in addition to the obvious nutritional benefits. Teleki was quite certain of this after watching meat-sharing episodes. He noted how strongly patterns of meat sharing depend on the hunter's social status. He and others argued that the social fabric of chimpanzee society is more complex than that of smaller-brained social hunters such as wolves and lions.[35] They ways in which meat is involved in social dominance and power are also more complex.

Some researchers have claimed that hunting is solely about energy acquisition. Gilby and his colleagues did not find compelling evidence during their study at Gombe that male chimpanzees hunt in order to share meat strategically.[36] This stands in contrast to other studies that have examined hunting and meat sharing. At Ngogo, Watts and Mitani concluded that male chimpanzees share meat preferentially with particular other males. Male bonding and male efforts to rise in rank by currying favor with the right higher-up were major motivating forces in obtaining meat in the first place. The researchers concluded, as I did, that while there is no single, simple explanation for hunting, nutritional and energetic explanations that dismiss social factors are likely to fall short.[37]

Chimpanzees engage in behaviors that are learned and transmitted culturally, from grooming postures to small symbolic gestures related to social status. Meat sharing is among these behaviors. The role of the individual matters greatly as well. Frodo was the most important and pervasive procurer and sharer of meat during the 1990s and 2000s at Gombe. In Gilby and colleagues' results on meat sharing, Frodo was involved in more than half the meat-sharing episodes. The authors opted to exclude him from the analysis, because his impact was so great that they felt it biased their overall hunting results.[38] This was an odd decision given how much we know about the importance of an avid hunter as a catalyst of the hunt. Great apes are different from other nonhuman animals precisely in the importance of individual life histories and personas. Wolf and lion societies no doubt have their dominant figures too, but the catalytic role that Frodo played in the hunting ecology of Gombe, which was likely related to his own high-ranking lineage, was a key to understanding hunting behavior in that community.

After years studying chimpanzee predatory behavior at Gombe, I wrote a series of papers with colleagues in which we showed that there were numerous factors that influenced hunting. These were not mutually exclusive, and they suggest the complexity of this fascinating behavior.

Forest structure. Chimpanzees are more avid and more successful hunters in areas of forest in which the tree canopy is broken due to tree falls or the presence of smaller tree species. This no doubt makes it harder for the colobus to either flee or mount a successful defense. It might even be that hunting is more frequent in the dry season at Gombe because visibility for the hunters is better without leaves on the trees. The structure of the forest is a proximate reason for hunting success but not a motivation for hunting itself.

Prey defense tactics. Nearly all research on hunting has focused on the chimpanzees as predators. Few studies have looked at the impact of hunting on the population biology of the prey species. I arrived at Gombe initially planning to spend more time watching red colobus than chimpanzees, in order to understand the factors that make a hunt succeed or fail from the perspective of the prey. I found that the larger the number of male colobus defending the group, the less likely the chimpanzees are to succeed in capturing one. But this only holds true up to a point. When there are five or more chimpanzees hunting together, there is not much the male colobus can do to limit their losses. In that scenario at least one colobus is usually killed.

Male chimpanzee demographics. Hunting patterns change through the years as the composition of hunting parties changes. Young male chimpanzees grow up, and older hunters die. Some males are avid, fearless hunters, while others are less so. In some generations there may be a cohort of five or six adult males who consistently hunt together; in other generations it may be eight or ten. Since the number of males in the hunting party strongly influences hunting success, a baby boom of males today can dramatically change hunting patterns as the males mature a decade and a half later. Watts and Mitani showed that male bonds were closely connected to the frequency of hunting. They argued that males' desire to share with male allies, presumably for political reasons, might be a force behind the desire to hunt.[39]

The number of male chimpanzees in the hunting party. A key determinant of hunting frequency and success is the number of males in a hunting party. At Gombe, a hunting party with more than five males will simply overwhelm the defensive counterattacks of male colobus. If larger hunting parties are more successful, then the factors promoting large party size will be predictors of hunting. Two key factors influence party size: the availability of fruit and the presence of female chimpanzees with sexual swellings. At Gombe, we found that the presence of females with swellings was the most important factor associated with many males hunting together. This was likely an indirect effect. The males congregate around a swollen female. These periods may last several days, and large parties often form. Large parties tend to hunt when they encounter prey, and they're highly successful because of the sheer number of males hunting together. So female reproductive cycles indirectly impact hunting patterns and also therefore impact the population biology of colobus monkeys.[40] The number of cycling females at any given time influences the mortality rate of colobus monkeys. This pattern held true through the 1990s at Gombe, although an analysis of an overlapping time period done by Gilby and other Gombe researchers a decade later did not find the same result.[41] As particular males matured and gained hunting prowess, and as others grew older and died, hunting frequency ebbed and flowed. The same applied to female chimpanzees, some of whom were wildly popular with males, leading to the formation of large parties around them, while others were almost ignored even when they possessed sexual swellings.

We would expect hunters to be more successful when they cooperate, just as human hunters are. But measuring cooperation is surprisingly difficult. We can't interview the chimpanzees to find out their underlying motivations as we would for human hunters, so researchers have used statistical tests that suggest cooperation. As the size of the hunting party increases, the amount of meat available per individual should decrease when a kill is made. If each male instead ends up with more meat as the size of the hunting party increases, that implies cooperation. If per capita meat consumption declines with increasing party size, it suggests a lack of cooperation.

Interpreting cooperative behavior is not, however, straightforward. When an increase in the number of hunters working together results in more meat for each of them, cooperation is suggested. This benchmark has been applied in studies of other social hunters such as lions and wolves. I estimated individual meat consumption to increase in larger hunting parties. I did not, however, see anything that looked like cooperative hunting, nor has any other primatologist who has studied hunting at Gombe. At Taï, Boesch inferred cooperation, even claiming the chimpanzees took on specific roles during the hunt—blockers, attackers, and more—that were crucial in catching the colobus prey.[42]

The political and social aspects of meat sharing. Anyone who has watched the aftermath of a chimpanzee hunt has been struck by the intensely political nature of the scene. Sharing is neither liberal nor random. It usually follows lines of kinship, nepotism, and opportunism. Females rarely hunt and are rarely successful when they do, yet they need the nutrients and calories from meat as much as males do, particularly when they are pregnant or lactating.

Of the dozen or so factors that I identified as strong influences on hunting, the one that stood out from the others statistically (not all were easily measurable) was the presence of one or more females with sexual swellings in the hunting party. This was also the one factor that provoked the most debate. It suggested that the presence of swollen, sexually available females might be an important incentive for males to hunt. We know that males use meat for a variety of political purposes, such as cementing alliances with other males or snubbing rivals. Meat sharing is a highly political and Machiavellian behavior.

In a series of scientific papers that I published on hunting, I theorized that one aspect of male manipulation of others was the use of meat to entice females to mate with them. Sex is not exactly in limited supply in chimpanzee society. There are often hundreds of matings for each conception. As we've seen, females are highly promiscuous and use sex to strengthen bonds with particular males to benefit themselves and perhaps protect their infants from male aggression. But a few more mating opportunities is always a good thing from a Darwinian perspective, especially for a lower-ranking male who might live under the thumb of the alpha. With a handful of colobus meat, and with the females present relishing meat but rarely capturing it themselves, the carcass becomes a magnet and a bargaining chip.

When I described this finding in the scientific literature and later in popular articles and books, there was a backlash on two fronts. First came criticism from primatologists who challenged my data and interpretations. Some of them argued that in such a promiscuous species, a few extra matings with a male would not be important enough of a payoff to incentivize his hunting behavior. I had argued that once a male has meat, he is Machiavellian in his sharing behavior; one such Machiavellian strategy involved female chimpanzees. Other colleagues analyzed larger data sets than my own and came to opposite conclusions from mine. This cast doubt over whether my conclusions based on an earlier, smaller data set held water.

More recently, however, the assertion that male chimpanzees may receive additional sex in exchange for meat has received renewed support. Christina Gomes and Boesch of the Max Planck Institute for Evolutionary Anthropology reported that male chimpanzees at Taï form mating associations and receive more sex from females with whom they share meat.[43] This finding was based on their analysis of a larger and longer-term data set than my own from Gombe. It makes sense; wild chimpanzees live several decades, and thousands of opportunities for sharing relationships develop. The potential for long-term exchanges of services between chimpanzees has been well demonstrated in captive settings. Gomes and Boesch extended the long-term nature of such exchanges to the wild. A second recent piece of evidence in support of my own findings came from my former doctoral student Robert O'Malley of the University of Maryland and colleagues, who found that female Gombe chimps with sexual swellings eat much more meat than females without swellings. Since females usually acquire meat by receiving it from males, it's clear that males do indeed share preferentially with swollen females.[44] These two recent studies don't necessarily revalidate my own findings, but they do make it clear that the picture is complex and the current explanations far from complete.

There were also criticisms from social scientists. Decades ago, anthropologists had generated a theory of human origins that revolved around hunting. Sherwood Washburn and Chet Lancaster had called it "Man the

Hunter." They argued that our big brains and intelligence were driven by the premium placed on cooperation and communication within men's hunting parties as they pursued big game.[45] But then why exactly did women's brains evolve? Later studies of the !Kung and other foraging people showed that the majority of animal protein was collected by women in the form of small animal prey. Men hunted large prey, but often unsuccessfully. Man the Hunter was promptly attacked as sexist, and rightly so. My articles and books described the largely male nature of chimpanzee hunts and the ways in which males share food with females in the name of political and sometimes sexual manipulation. These accounts evoked memories of Man the Hunter in some anthropological circles. Some of the media covering my work interpreted my finding that male chimpanzees sometimes exchange meat for sex as though I had uncovered the origins of prostitution.

While there have now been a few studies of the predator-prey relationship between chimpanzees and red colobus, we still do not know the exact role of chimpanzees as predators. Are they capable of driving colobus monkeys to local extinction? Research at Gombe suggests not, while work at Ngogo suggests it's a possibility. With more data, we should be able to answer questions such as, what are the long-term nutritional benefits of eating meat to chimpanzees? Are the benefits mainly political or purely energetic? Is hunting cooperative, or does it just appear that way to some scientists?

Chimpanzees, perhaps more so than any other nonhuman animal, are individuals. There are politically driven males who strive for dominance their entire lives. Other males seem uninterested in status. Females can be caring, conscientious mothers or carefree, incompetent mothers. These personality differences can translate into populational differences. This makes chimpanzees endlessly fascinating, but it also makes it difficult to generalize about them. The likeliest reason that researchers studying the same behaviors in different sites arrive at different results is that chimpanzee communities have well-documented variation in traditions, which include styles and tactics of hunting and meat sharing. While some of this variation may be attributed to some physical or energetic variable, some of the differences are clearly learned. Even the factors at play in the same chimpanzee community will change over the course of a few years, making comparisons of hunting behavior in the same forest between different decades difficult to interpret. Over the past half century, we've learned not to speak of "The Chimpanzee" as though individuals or populations are interchangeable. Today we regard that idea as antiquated and quaint. This is also true of the overgeneralization of the reasons for hunting behavior among the many chimp populations we know of.

We have no reason to believe that meat is essential to the health and well-being of wild chimps. It's a wonderful package of calories, fat, protein, and essential amino acids to supplement an otherwise largely plant food diet. Hunting is the only way the chimpanzees can obtain those nutrients in a single package. Beyond a nutritional supplement, meat is yet another way in which these big-brained apes manipulate one another and display their social intelligence. As chimpanzee researchers compile abundant geographic variation in behavior and cultural diversity, it may turn out that the variation in hunting and meat-sharing patterns has multiple explanations. As in traditional human societies, meat represents one thing to females: calories and nutrients for offspring and fetuses. It may represent something very different to males: an opportunity to enhance status and sexual access while getting nutrients as a bonus. The various patterns seen in the many research sites are responses to local ecologies, local demographics, and local cultures. While these make a universal explanation unlikely, they reinforce the fact that hunting and meat sharing provide us with key clues about the seeds of our own diet, and our intellect.

CHAPTER 8

Got Culture?

WHEN SCIENTISTS BEGIN studying a new community of chimpanzees, they must first spend years gradually making those animals accustomed to their presence. It's easier to habituate chimpanzees to humans when there are other chimpanzees who are already accustomed to people in their midst. We often see this when females from a habituated community, already accustomed to people, have transferred into a new community. The fearful, inexperienced animals take note of the lack of concern by the human-savvy ones and calm down. This is social learning, and it's at the heart of chimpanzee culture.

Every primatologist I know accepts that cultural behavior in chimpanzees is real. But I have many colleagues in other fields of anthropology and psychology who don't accept a role for social learning in many chimpanzee behaviors, from hunting to food sharing, nor in geographic variation in community-wide behaviors. In fact, at least two of the papers cited in the bibliography of this book are authored by scholars who explicitly place the word "culture" in quotation marks. This reluctance to embrace chimpanzees as cultural animals may be healthy scientific skepticism. Or perhaps it reflects the opposite, a certain obtuseness about chimpanzee uniqueness. In my view—and in the view of nearly all my colleagues—the evidence is clear. Culture is social learning writ large, and it plays a major role in chimpanzee life.

It's not easy to define culture even when we're talking about humans. Culture is a human universal, so anything that we do that is not genetically hardwired could be called cultural behavior. I have been scolded by cultural anthropologists for referring to chimpanzee behavioral variation as cultural diversity. William McGrew took famed ethnographer Alfred

Kroeber's time-honored checklist for human culture and found many chimpanzee studies that fitted the same categories.[1]

Cultural anthropologists are loath to grant culture to chimpanzees because their own definitions of culture are exclusive rather than inclusive. They see symbolic behavior as the heart of culture. That is, humans *create* things we call cultural artifacts. They invent symbolic sounds called words that have no connection to the ideas they represent. Nothing in the word "red" tells you what something red looks like. And the anthropologists are right—language is a central aspect of human culture that goes far beyond anything that chimpanzees do in the wild or can be taught to do in captivity. But this shouldn't deny them culture.

Culture is socially learned behavior. Over time the learned behaviors become entrenched as local traditions. These traditions spread through the whole community until we can characterize a given chimpanzee community as stone tool using or termite probe fishing. Particular traditions, such as one's posture while grooming a group-mate or while holding an ant-fishing probe, develop their own *style*. We don't often use the word "style" in primate behavior, but a behavior pattern that becomes a trend is exactly that. It may last a long time or briefly; it may spread far and wide or stay contained within one community or even one matriline. Richard Wrangham and colleagues recently showed that at Kanyawara, chimps employ lifelong grooming techniques learned from their mothers rather than adopt other styles they may see around them in the community.[2] Such learned behaviors are no longer idiosyncrasies; they're group traditions. This, McGrew argues, is why variation in behaviors among chimpanzee communities fits the human culture definition, albeit on a simpler scale.

Most cultural behaviors in chimpanzees seem quite trivial. Two Mahale chimpanzees clasp their right hands above their heads while their left hands groom their partners. At Gombe, the grooming partners grasp branches instead of hands. In some forests, chimpanzees use small sticks to probe into tree bark for ants; in other forests they don't, even though the same ants are readily available. As the number of long-term field studies of chimpanzees grew, so did our understanding of the scope of cultural diversity across Africa. Well into the fourth decade of chimpanzee field research, Andrew Whiten compiled the full complement of cultural variation across Africa, using information contributed by coauthors from each of the seven longest-term studies. They identified thirty-nine behaviors across seven sites that appeared to be culturally and not ecologically induced. These included both foraging and social traditions.[3] They also included leaf grooming, one of the few symbolic cultural traditions that varies from site to site. It's practiced by many East African chimpanzee communities. The chimps pluck leaves and groom them as intently as they might pick through the hair of another chimp. Other chimps understand the behavior to be a signal of a desire to groom or to be groomed. Leaf clipping is another tradition that indicates symbolic communication. A male chimpanzee audibly clips the leaves from a plant stem using his fingers and teeth. It appears to signal a desire for sex with a particular female, and perhaps some sexual frustration.

How do such traditions begin and spread? Primatologists are still engaged in a half-century search for the social mechanisms by which new traditions emerge and become entrenched. We don't really know yet whether such traditions appear infrequently and are then readily adopted by others or pop up often but are rarely adopted. It's even possible that the pattern of traditions across Africa may reflect the selective random local extinction of traditions, as Toshisada Nishida and his colleagues have reported. Culture may emerge and spread within a community readily but disappear just as often.[4]

Researchers at the University of St. Andrews have done some of the best recent work on the emergence and transmission of tool use among wild chimpanzees. Catherine Hobaiter and colleagues used social network analysis to understand social transmission in the Sonso community at Budongo. They observed a new behavior not previously seen in two decades of research. The alpha male made a sponge of moss that he had gathered from a tree trunk, and used it to soak up water from a large puddle. He did this in the company of the highest-ranking female, who watched him intently. Over the following week, that female and at least six other members of the community made moss sponges and used them at the same waterhole. The rest of the community drank using their mouths or with the more time-honored leaf sponges. Hobaiter and her colleagues estimated that nearly all the observed moss-sponge use occurred through social transmission and not individual learning. This is not a trivial distinction. In some cases of apparent social transmission, careful study has shown that the practitioners of a tradition may have learned the behavior on their own independently. This marked the first time that both tool use and its mode of transmission had been documented in the wild.[5]

There has been a tidal wave of awareness of cultural behavior since the late 1990s. As more long-term studies accumulate, we see new evidence of innovations and transmission. As researchers locate new field sites, they also locate new chimpanzee cultures, each with its own unique set of traditions. For chimpanzees that are too shy to be watched, we now use remote camera traps. These are durable, motion-activated still and video cameras that can be placed along trails in forests where chimpanzees are utterly unaccustomed to people. When we retrieve the memory cards, we find recordings of chimpanzees engaging in behaviors never before seen, including new forms of tool use. These observations would have required many years of habituation to see in person.

Tools

Although grooming styles are fascinating aspects of local cultural diversity, most of the evidence for chimpanzee culture comes from technology. There is no debate about chimpanzees' being the most proficient tool users among all nonhuman animals. Their closest kin, the bonobo, for all its much-vaunted social sophistication, doesn't begin to compare in technological sophistication to chimps. In the wild, most chimpanzee tools are used to procure food. What limited tool use exists in wild bonobos is seen more in nonfeeding contexts: sticks used in play or leaves used as rain umbrellas. Thibaud Gruber and his colleagues of the University of St. Andrews compared tool use among captive bonobos and chimpanzees and found no difference in proficiency or diversity. The main distinction between the two apes was that bonobos used tools primarily in a play context, while chimpanzees utilized them more widely. In both species, females were more avid tool users.[6]

Chimpanzees use tools mostly for acquiring plants, honey, and insects that are difficult to obtain by hand. In rare cases they may also use tools to obtain meat. Let's consider plant foods first. Most of the chimpanzee diet is fruit, and foraging for fruit takes up much of a chimpanzee's waking hours. Once a tree laden with ripe fruit is found, the chimpanzees eat their fill. But sometimes the best part of the fruit is hidden inside a shell that is too hard for a chimpanzee's jaws to break open. In East

Forms of tool use discovered since 2000 among wild chimpanzees.

Tool Use	Site	Reference
Digging stick	Ugalla, Tanzania	Hernandez-Aguilar, Moore, and Pickering 2007
Ant-digging stick	Ngel Nyaki, Nigeria	Dutton and Chapman 2015
	Gashaka, Nigeria	Fowler and Sommer 2007
Earth-perforating stick	Goualougo, Republic of Congo	Sanz, Morgan, and Gulick 2004
Army-ant-dipping tool set	Goualougo, Republic of Congo	Sanz, Schöning, and Morgan 2010
	Kalinzu, Uganda	Hashimoto et al. 2015
Fluid-dipping tool set	Ngogo, Budongo, Uganda	Gruber et al. 2009
Honey tool set	Bulindi, Uganda	McLennan 2011
	Ngotto, Central African Republic	Hicks, Fouts, and Fouts 2005
	Loango, Gabon	Boesch, Head, and Robbins 2009
	Moukalaba-Doudou, Gabon	Wilfried and Yamagiwa 2014
Hunting stick	Fongoli, Senegal	Pruetz et al. 2015
Pounding/grating stone	Ngel Nyaki, Nigeria	Dutton and Chapman 2015
Stone throwing/piles	Many sites	Kühl et al. 2016

Africa, that renders the food inedible, although in a few cases the chimpanzee will pound the fruit against a rock or a tree trunk in hopes of cracking it. In West Africa, chimpanzees have invented stone and wood hammers to solve the problem. Trees at Taï provide an enormous bounty of calories for several months of the year. In the late 1970s, Christophe Boesch documented the use of stone or wood hammers by Taï chimpanzees. This had first been reported a century earlier, but now a modern scientist had confirmed it. At the base of several species of fruit trees that drop hard-shelled nuts, the chimps gathered to create workshops. A tool user carefully places a nut into a depression worn into the surface root of the tree—the anvil—and then cracks it open with a hammer of stone or wood.[7] This exciting finding was much discussed during the 1980s.

Chimpanzees in East Africa, presented with an abundance of stones from which to create hammers, have never invented this technology. Nor do all West African chimpanzees use hammers in foraging.

Recent studies of nut-cracking chimpanzees in West Africa have taken our understanding of the behavior to new levels. Giulia Sirianni and her collaborators at the Max Planck Institute for Evolutionary Anthropology examined how Taï chimpanzees select their tools. They hypothesized that tool selection should be based on energy efficiency for the task at hand, in order to maximize power and control of the tool selected. Observers have watched for more than three decades as Taï chimpanzees approach the job of nut cracking with circumspection, then heft a tool that seems right for the task. There's a lot of cognitive processing in that decision, based on experience and judgment.[8] While chimpanzees at another site (Bossou) use stone hammers provided for them experimentally, the Taï chimps use tools naturally. And Adrían Arroyo of University College, London, and colleagues at Kyoto University showed that the patterns of abrasion on stone tools made by captive chimpanzees match those on stone tools used by ancient humans.[9]

Physical properties, in terms of size and weight, ought to be of paramount importance in choosing a tool. Sirianni and her colleagues found that Taï chimpanzees took into account a number of variables when selecting a tool. They chose stone hammers over wooden clubs and preferred hard wood to softer wood when stones couldn't be found. Christophe Boesch and Hedwige Boesch-Achermann had reported in the early 1980s that Taï chimpanzees select their hammers and anvils depending on the nut; wooden hammers were mainly for soft *Coula* nuts, while stone hammers were chosen for the harder-to-crack *Panda* nuts.[10] Stone tools tended to be heavy, whereas wooden clubs were lighter in weight. They selected lightweight hammers when the "anvil" was the tree itself, and heavier hammers when they were closer to the anvil (perhaps to avoid carrying heavy stones longer distances). The sequence in which tools were selected suggested a mind that was thinking several steps in advance about what would be needed to accomplish the task of cracking the nut. The thought process appeared to take into account both use of the tool and transport of it to the site where it was to be used.

Nut cracking begins early in life. Taï chimpanzees start in infancy but take up to four years to master the technique of cracking open a *Coula* nut. At Bossou, Tetsuro Matsuzawa of Kyoto University found infants trying to crack oil-palm nuts as early as three and a half years old, but they were rarely proficient at the task before the age of eight.[11] Noriko Inoue-Nakamura, also of Kyoto University, found that infants had learned the basic task of nut cracking by age two and a half but took years longer to master the sequence of steps needed to actually open the nut. This was partly a function of trial-and-error learning and partly of the amount of time infants spent in object manipulation activity. By age four, the cognitive process had developed to the point that the infant employed all five of the necessary steps: pick up the nut, place it on the anvil, hold it there, strike it, and eat it. At age two, infants had learned only two or three of the steps, and they often performed them out of order.[12] Active teaching has rarely been seen in wild chimpanzees: the infants learn by watching and then attempting, and there is no reinforcement or reward from the mother for a job well done. There are exceptions that offer exciting insights into the nature of learning, and of teaching. At Taï, mothers sometimes manually guide their offspring's nut-cracking technique until they get it right. In the Goualougo Triangle, Stephanie Musgrave of Washington University found that termite probers donate their sticks to learners.[13] Such technology transfer is a high form of learning, and is reminiscent of what parents do with children when teaching them to eat with utensils or hold a pen.

Cornelia Schrauf of the University of Vienna and her colleagues studied the acquisition of nut-cracking skills in captivity. They found that chimpanzees made the quite logical decision to select stone hammers that were heavy, presumably to minimize the number of times the nut had to be struck before it cracked. Since the hammers all looked more or less alike, the nutcracker appeared to choose his tool based on weight, hefting each candidate stone in his hand. Unlike in the Taï field studies, weight alone dictated the choice of tools. The most experienced tool users were, with one exception, the most proficient and efficient at hammer use.[14]

Chimpanzee Archaeology

Research in the 1970s on stone tool use by West African chimpanzees was wildly exciting to scientists studying human evolution. Archaeologists had been trying to understand the origins of stone tool use by early humans, and suddenly a rather early hominin-like ape was actually using them in the present day. In the 1990s, prehistoric archaeologists Nicholas Toth and Kathy Schick of Indiana University began a collaboration with Sue Savage-Rumbaugh and Duane Rumbaugh of Georgia State University to study stone tool selection and use by the most famous bonobo who's ever lived: Kanzi. Kanzi was better known for his linguistic prowess, but he also proved to be a very competent tool user, learning to use stone tools provided him to cut ropes in order to obtain treats. His ability to learn to use stone tools came from his ability to imitate humans who showed him what they could be used for.[15]

There has been a long debate about the cognitive abilities of captive chimpanzees. One school of thought, to which psychologist Michael Tomasello of the Max Planck Institute for Evolutionary Anthropology and others belong, is that chimpanzees do not grasp the concept of imitation, at least not in the human sense. Children understand that copying a task means copying each step in a sequence to achieve a goal. Chimpanzees are able to understand the goal—obtaining a food treat, for example—but cannot grasp the concept of copying each step in the process they are asked to learn. Instead, according to Tomasello, chimpanzees emulate, achieving the same goal without the same sequence of steps.[16] Kanzi exemplified this in figuring out novel ways to use a sharpened stone tool to obtain food. He initially smashed a rock against the floor or walls of his enclosure to create a sharp cutting edge, instead of using a second rock to flake off an edge, as early humans would have done and the experimenters hoped Kanzi would too. In time, however, he learned to be a proficient tool manufacturer and user.

Meanwhile, archaeologists turned their attention to the natural use of stone tools by the Taï chimpanzees. The nut-cracking chimpanzees gather stones and bring them to the base of the trees where they are to be used. In doing so, they create accumulations of stone tools that, over time, come to resemble human-created archaeological sites. Julio Mercader of the University of Calgary and his collaborators examined the archaeological context of stone tools at Taï, effectively creating the new academic discipline of ape archaeology. Excavations revealed stone accumulations dating to over four thousand years ago. Chimpanzee stone tools are not flaked or modified in the way that even the most primitive human stone tools are. They exhibit the wear-and-tear marks of repeated striking.

Their accumulation pattern on the landscape is also diagnostic of a behavioral rather than a natural stone deposit. And the stone tools themselves revealed something of their purpose. Starch granules were detectable on the ancient stones, consistent with the residue that would be left by a tool user striking them against the nuts for which Taï chimpanzees employ stone technology. The evidence was strong that Taï chimpanzees had been using stone tools for hundreds of generations. The idea that chimpanzee prehistory mirrors the technological prehistory of the earliest humans is as intriguing a finding as any in chimpanzee research in recent years.[17]

As exciting as recent studies of stone tools are, they are hardly the only new discoveries about chimpanzee technology. Crickette Sanz and her colleagues reported that chimpanzees use branches as spades by placing their feet on them and driving the end into a termite mound to perforate it the way a gardener uses a shovel to break ground.[18] R. Adriana Hernandez-Aguilar and her coauthors described chimpanzees in Ugalla, Tanzania—one of the most arid places that chimps have ever been studied—using digging sticks to reach tubers growing underground. During the rainy season, Ugalla chimpanzees use tools to get at these nutritious items that they otherwise would not be able to reach.[19] Tubers are believed to have been important foods for early humans as well, because they survive drought, and there is suggestive evidence that early hominins foraged for them. Modern hunter-gatherers also use digging sticks to extract wild tubers from the ground. Chimpanzees have also been reported to use digging sticks to reach underground ant nests in Gashaka Gumti National Park, Nigeria, and elsewhere in West and central Africa.[20] Unlike stone tools, digging sticks would not have fossilized, so understanding their use by modern ape analogy is the best we can hope for. This has made the discovery of chimpanzee digging sticks as exciting as that of stone tools.

Honey and Insect Tools

Stone tools and digging sticks are for reaching otherwise unavailable plant foods. But chimpanzees also crave animal protein, as well as the other nutrients that animals can provide. All across Africa, chimpanzees collect honey from African honeybee nests. Usually, they employ a purse snatcher's approach to getting it: smash the comb in a tree, grab some of the sweet stuff, and run to avoid the bees. But for the nests of other species of bees, complex tool kits are used. In Bwindi Impenetrable National Park, near bee nests we found stick tools of varying diameters and lengths, which were redolent of honey. At the base of trees that held honeybee hives in the trunk far above, we found large sticks, which were apparently used to break into the comb. A long stick would allow them to break the hive open while keeping a bit of distance between the attacking bees and themselves. Bwindi chimpanzees also foraged for tiny stingless bees, whose nests were usually underground. Here they used small sticks, perhaps because the task at hand requires more delicacy and there is no need to maintain a distance from the bees.[21]

Honey tools have been known for decades, but recent years have seen increasing reports of composite tool sets for gathering honey, as well as insects. For example, Matthew McLennan of Oxford Brookes University reported the use of tool sets for obtaining honey in Bulindi, a forest near Budongo in western Uganda. Bulindi chimpanzees used digging sticks to break into underground stingless bee nests, then switched to using small, flexible sticks as delicate probes to reach the bee nest tubes in which the honey was located. This tool set, which had previously been found widely in central Africa, may also be in use at Bwindi and elsewhere in East Africa.[22] Similar evidence has been reported by Thurston Hicks of Central Washington University and his colleagues in Ngotto Forest in the Central African Republic.[23] McLennan concluded that honey represents a nutritious, high-carbohydrate treat rather than a staple or a fallback food, but it's highly sought after when available.[24] When we expand the geographic range of a tool practice, we effectively are documenting a new culture.

The use of complex tool sets to get honey has also been documented in Gabon in Loango National Park, by Boesch and his colleagues. Loango chimpanzees used three- to five-element tool sets, which would be among the most complex composites of technology ever recorded in a nonhuman animal. They extracted honey using a sequence of tools in which a given tool was only useful because a different tool used earlier paved the way for it. The names for these tools—"pounder," "enlarger," "collector," and "perforator"—are self-explanatory. While the overall number of tools found—more than a hundred at one tree—was larger for extracting honeybee honey, the diversity and complexity of the tool set was greatest for stingless-bee honey. Loango chimpanzees thus employ the same sort of hierarchy of logical steps to achieve a honey-extraction goal using technology that Taï chimps use for opening nuts.[25] These observations build on others made in central and West Africa over the past fifteen years. They indicate that tool composites are more widely used in these regions than in East Africa, for unknown reasons. Gruber and colleagues argue that the evidence is overwhelming that honey tool use is acquired through social learning, not individual learning based on ecological variation.[26] It would stretch the imagination even further to argue that chimpanzee tool use varies regionally due to "tool use genes." As is the case with East African chimpanzees that have access to plenty of stones but don't use stone hammer tools, the reasons for the lack of innovation of sophisticated honey tool sets in East Africa may be purely cultural innovation, or perhaps cultural extinction.

No observation made by Jane Goodall in the early 1960s changed our view of ourselves and other primates more than that of tool use. Early in her study, she observed chimpanzees fashioning probes from twigs. They poked them into tunnels of termite mounds and delicately withdrew the sticks with soldier termites clamped by their mandibles onto them. Nearly sixty years later, we have a deeper understanding of "termite fishing" and of foraging behavior for ants and other insects. Termite fishing is not limited to East Africa in the way that stone tools are limited to West Africa. Stephanie Bogart of the University of Southern California and Jill Pruetz documented chimpanzees using termite-fishing probes at Fongoli, Senegal, one of the westernmost locations at which chimpanzees have been studied.[27]

The tool kit of a given chimpanzee community is complex enough that it takes years of practice to be a proficient tool user. The process of learning these technological skills has been studied in great detail in the past fifteen years. At Gombe, Elizabeth Lonsdorf studied the development of termite fishing among young chimpanzees. Each infant must learn a hierarchical sequence of skills, just as Taï nut-cracking chimpanzees do. Gombe chimpanzees must first learn to locate and identify a tunnel in which there are likely to be large numbers of termites. Making the tool itself is central to the task, followed by the technique involved in inserting the tool into the tunnel and extracting termites. Having tried this myself, I can say it's not difficult with an adult's manual dexterity—the hardest part is identifying which termite mounds and tunnels are termite-rich enough to be worth the trouble of fishing them. But the skill set, simple enough for a primatologist, takes years for an infant chimp to acquire.[28]

Lonsdorf found that Gombe females developed proficient termite-fishing skills at a much earlier age—sometimes a full year earlier—than their male counterparts. They began the learning process earlier too, watching their mothers intently and gaining basic competence more than two years earlier than males. Females were ultimately better tool users than males. Most interesting, females appeared to learn and copy their termiting techniques from their mothers, while males did not. It's clear from Lonsdorf's and earlier studies that females are more avid and skilled at termite fishing than males are.[29] They may also learn the process somewhat differently from the way males do, given the sex differences in tool use. Female chimpanzees appear to be simply better at tool use than males are. When the first rains of the year arrive, the termite mounds soften and males turn their attention to termiting in earnest. Females, meanwhile, have been intently foraging for the insects all year long. Does this suggest a pattern in human origins in which females may have led the way in the advent of technology? It raises the possibility that some sex differences in human ancestry, and by extension in modern humans, might be more hardwired than learned.

In the early days of primate field research on diet, we simply observed what our study subjects ate. The percentage of the diet that was leaves, fruit, and such took center stage. It took years before anyone realized the importance of studying the plant foods themselves to see why certain species were chosen during certain seasons or in certain stages of ripeness. That was no easy task, but today we have a rich understanding of the nutritional biology of plant foods. It likewise took researchers decades after the dawn of tool research to begin to examine why certain insects are consumed and others are ignored. This is because measuring food intake and the caloric and nutrient value of the range of foods in a complex tropical forest is truly daunting.

A few studies have made inroads into understanding the nutritional value of the animal foods that chimpanzees consume. Robert O'Malley and Michael Power, the latter of the National Zoological Park, analyzed the nutritional content of insects consumed at Gombe and also those insect species that were not eaten. They hypothesized that Gombe chimps would select from the diverse array of insects living in their habitat the insect prey that maximize their intake of energy involving fat, protein, fiber, and ash. Insect exoskeletons are composed largely of chitin—a carbohydrate—but its relative indigestibility makes it the animal equivalent of plant fiber. O'Malley and Power analyzed samples of the insects living in Gombe and compared their nutrient content to that of insects that the chimpanzees actually eat. They found that insect prey had a significantly higher fat content per gram of body weight than insects that were not eaten. On a per-insect or per-termite-stick-dip basis, insect prey were higher in fat, protein, and utilizable energy than insects not preyed on. Measured per gram, the nutritional content of termites, ants, and other insects eaten by Gombe chimpanzees was comparable to that of the meat of the monkeys, antelope, and pigs that chimpanzees also eat. Chimps eat a number of insect species that are high in energy value, but they also ignore abundant species that contain readily available nutrients. O'Malley and Power's work showed that insect eating is not just a snack; it can be an important component of the diet.[30] As McGrew has long argued, insects can be at least as valuable a food item as much more ballyhooed meat.[31]

Termites are thus not the only insects that chimpanzees relish. In the course of honey collecting, they ingest the larvae of bees. And they consume a diversity of other insects, some of them collected on purpose and many simply swallowed along with leaves or fruit. Ants are also an important food source. Anting can be similar to termite fishing; some ants live in nests and can be probed gently out of holes in trees or from under bark. But one of the main forms of ant nutrition also offers an important window into the mind of the chimpanzee. Army ants (also called safari ants) of the ant genus *Dorylus* are famous for their large nests of upwards of ten million members; their columns flow like spilled chocolate across roads and snake through forests as the colony moves from one nest to another. They're infamous for their tendency to swarm, during which they attack and eat everything in their path, from other insects to lizards and frogs. Chickens locked in a henhouse when an army ant swarm arrives will suffer the same awful fate. There are few experiences as unpleasant as awakening in the middle of the night in a tent, hearing a scratching sound all around, and feeling something biting your scalp, then switching on your flashlight to reveal that the walls of the tent are alive with the bodies of the marauding swarm.

Army ants, despite their aggressiveness, are much-desired food items for chimpanzees in many (but not all) forests. The genus *Dorylus* is on the chimpanzee menu at Gombe (Tanzania), Taï (Ivory Coast), Gashaka (Nigeria), and Nimba (Guinea). They're present but apparently not eaten at Mahale (Tanzania), Budongo (Uganda), Kibale (Uganda), or Lopé (Gabon).[32] The ants present hungry chimpanzees with two problems. First, *Dorylus* soldiers are armed with massive slicing mandibles that plunge into a victim and don't let go. Chimpanzees do their best to avoid them, and they accept some bites as a small price to pay for all the nutrients the ants contain. The second problem is that *Dorylus* nests, unlike earthen termite mounds, move around. So the ants can't serve as a dietary staple, because the chimpanzees can't rely on finding them regularly.

When a chimpanzee locates an army ant nest, she takes a skillful and careful approach to making a meal of it. A stick, more substantial and longer than the slender probes used for termite fishing, is selected. There's a functional reason for a long stick: it keeps the angry soldiers at arm's length. At some sites longer sticks are associated with the most aggressive *Dorylus* species. There is also cultural variation in the tools selected. Kathelijne Koops and her colleagues recently found that ant-dipping wands at Kalinzu Forest in Uganda varied in size among nearby communities despite identical *Dorylus* species in both sites.[33] A survey by Caspar Schöning of the University of Copenhagen and colleagues compared army-ant-dipping wands at fourteen sites across Africa and found that the aggressiveness of the ant species doesn't necessarily predict wand length, leaving room for purely cultural influences.[34]

Once the wand is chosen, the user suspends herself over the nest with legs and one arm splayed on branches above while the free hand does the dipping work. The wand is lowered into the chocolate-colored mass of the ant nest, then withdrawn and run through the other fist. The fist swipes off hundreds of biting ants, which are shoved quickly into the tool user's mouth and crunched up before too many bites to the tongue and lips can be inflicted.

Dorylus are the most well-known case of chimpanzee predation on ants, but they aren't the only important indicators of learning and culture.

At Gombe, the habituation and observation of a new chimpanzee community allowed us a glimpse into how cultural behavior is transmitted from one community to the next. More than thirty years after Goodall habituated the Kasekela chimpanzees, research began on the Mitumba community—a small community of fewer than twenty-five chimps—just to the north. O'Malley and collaborators documented the emergence of ant fishing in the Kasekela community at Gombe. The *Camponotus* ants are "fished" with stick probes from under tree bark and inside tree holes avidly at Mahale (where chimpanzees ignore army ants) and elsewhere across Africa. Ant fishing had never been documented in the Kasekela community from the early 1960s to the mid-1990s. Then, in 1994, ant fishing appeared and became a part of the repertoire of the main study community. It's possible the behavior had been occurring all along but had been missed by researchers because it was infrequent. Or perhaps the ants themselves had become more numerous and the chimpanzees' desire for them hadn't changed at all. Maybe a member of the Kasekela community had in fact innovated ant fishing on his or her own, and it had then been observed by the others and spread.

O'Malley and colleagues found evidence that another ant species long considered abundant, *Oecophylla longinoda,* may have declined, or at least that chimp consumption of them had declined. This might have led the chimps to turn to other ant species, which may have become more common in the absence of the competitor species. They also considered the possibility that a native member of the Kasekela community, Flossi (Fifi's daughter), may have invented the behavior and then watched it spread to her group-mates.

There was one more possibility: that ant fishing had been transmitted culturally from outside the Kasekela community via an immigrant. Before 1994, when Flossi's ant fishing was first observed, at least one female chimpanzee had immigrated from the Mitumba community to the north, bringing with her the ant-fishing culture of the Mitumbans. Ant fishing had been observed in Mitumba since the earliest days of observation of that community in the late 1980s. The Mitumba immigrant in question, Trezia, was a young adult female who may have been an avid ant fisher there before her immigration. Trezia may have brought the new tradition to Kasekela. This is likely an example of the way in which culture spreads in chimpanzees, and, by extension, in humans past and present.[35]

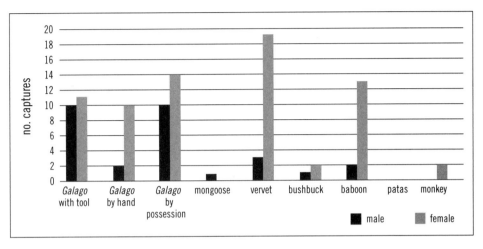

Hunting with and without weapons at Fongoli: sex differences and prey choice.

The last form of tool use intended to assist in procuring food is for hunting. Until recently, chimpanzees had very rarely been observed hunting with weapons. A predation by a chimp on a squirrel in Mahale in which the hunter, a young female, used a branch to force the squirrel from its refuge in a tree was the only clear-cut example.[36] This all changed with the observations by Pruetz and her colleagues of tool-assisted hunting at Fongoli in Senegal. The researchers have now witnessed hundreds of instances in which a chimpanzee uses a large stick, chewed to a point at one end, to injure or kill a galago (bush baby) holed up in a tree cavity. Both males and females, adults and juveniles, engage in weapon use, although adults are more successful than younger hunters.[37] Notably, females account for a higher percentage of the hunters than at other chimpanzee study sites. Pruetz argues that weapons allow chimpanzees who would otherwise be less able to run down prey in an open, grass-and-forest environment such as Fongoli to hunt with a high odds of success.[38]

Tools and Handedness

For as long as anthropologists have been writing about tool use, they've speculated about the origins of right- and left-handedness. Handedness is thought to be among the crucial steps to the use of tools, and since it involves localized brain function—hand preference being strongly linked to brain hemisphere dominance—it may have evolved in association with speech and language. Human populations are universally 80 to 90 percent right-handed, although the mix of genetics and learning that makes someone left- or right-handed is still poorly understood. Wild chimpanzees also exhibit hand use preferences, but such preferences in most tasks seem to be random and not population-wide, as they are in our own species (although captive chimpanzee populations have long been known to display group-level handedness). But recent research has shown that wild chimpanzees do exhibit striking laterality in hand use when using tools. Handedness studies have most often surveyed hand use preference across populations, but Sanz and colleagues have also shown strong right-handedness by termiting chimps in the Goualougo Triangle. Hand preference is strongest in the tool use tasks that appear to require the most manual dexterity and fine motor control, such a dipping sticks for ants.[39]

Lonsdorf and William Hopkins, the latter of the Yerkes National Primate Research Center, surveyed handedness among Gombe chimpanzees and found strong evidence for population-level hand use. They showed that the degree and direction of handedness varied depending on the task. They also found that handedness tends to follow family lines, as it does in humans. There was population-wide left hand preference for termite fishing. Population-wide right-handedness has been reported for nut-cracking chimpanzees at Bossou, and for leaf sponge use at Gombe. No sex difference in hand preference was obvious in the studies.[40] This task-specific hand use preference suggests that the demands of various tasks may select first for handedness and second for a particular hand, just as some types of human tool use are more easily done with one hand than the other. Captive studies in chimpanzee sanctuaries and laboratories have also found that the complexity of the task may select for strong hand use preference, as you'd expect. The harder the manual skill, the more specialization is favored for one hand over the other.

At Bossou, Tatyana Humle of the University of Kent and Tetsuro Matsuzawa examined hand use preference and found that older chimpanzees began to display more marked handedness. The strongest handedness was seen in the most complex tool use—nut cracking—in which the action of both hands needs to be coordinated. The least handed behavior was pestle pounding, in which the chimp uses a palm frond to pound the center of a palm crown until softened for better ingestibility. This also required both hands and because it appeared to be a tiresome task, the use of both hands was likely beneficial.[41]

Throwing is a behavior that chimpanzees display fairly often, but rarely in contexts that lend themselves to the kinds of interpretations about human origins in which human evolutionary scholars love to indulge. As a party of Gombe chimps approaches a rushing stream near a waterfall, their hair stands on end. Perhaps excited by the sound and sensation of the force of falling water, they rush forward and cross the rocks that line the stream. As they cross, they may grab a rock and hurl it, as a softball pitcher windmills a fastball. Researchers stand back and watch for flying objects. Chimpanzees also throw objects in excitement at other times and places. Hjalmar Kühl of the Max Planck Institute for Evolutionary Anthropology recently surveyed the use of throwing stones, and in particular the accumulation of such stones on the forest floor. This was in effect a different kind of chimpanzee archaeology. Kühl and her collaborators found something startling when they surveyed seventeen study sites across Africa and also compiled video footage of stone throwing. They found four communities in West Africa in which chimpanzees throw rocks against tree trunks or hold rocks in their hands while banging the trunks. Sometimes the rocks were thrown at a hole in the tree or an exposed gap in the trunk and landed inside it.[42] There was no obvious reason for this behavior; it wasn't related to foraging for food or impressing other chimpanzees.

The result of this stone throwing was accumulations of rocks inside or against the base of the trunks of trees. These rock piles resemble cairns left by hikers along trails. The most notable aspect of the stone throwing is that it appears to be a ritualized behavior. That is, the stones are thrown in a directed path at particular trees and left to accumulate there. How long this has been going on isn't known, but it's possible we're looking at a century of stone throwing for some of the largest and oldest trees that were targeted. The most obvious reason for the behavior would be that it's done by males in the course of their displays. The sound and sight of a rock slamming against a tree trunk would be an impressive way to enhance the usual male display of drumming with feet and hands on buttressed tree roots. Unfortunately, Kühl and colleagues don't report whether the stone throwing is a male behavior or is performed by both sexes.⁴³ And, most unfortunately, the media took some of the speculations expressed in the paper by the authors and ran with them, ascribing to chimpanzees not merely ritual behavior but "proto-religious" behavior. The idea that stone throwing is somehow spiritually motivated and the trees at which rocks are thrown are therefore "sacred sites"—as more than one media outlet suggested—is one of the sillier conclusions drawn from solid scientific research on animals in a very long time.

Tools, Culture, and Geography

I opened this chapter by considering the meaning of culture and how it may function in chimpanzee society. We feel sure the variation in many behaviors among chimpanzee communities across Africa is due to culture and not ecology, because the behaviors often don't conform well to the environment. Gombe chimpanzees have stones at arm's reach all day long but don't use stone tools. Taï chimpanzees gather and carry stones in a rain forest where stones are less available. The notion that each minor detail in cultural behavior might actually be linked to genetic differences among chimpanzee populations seems extremely unlikely. Stephen Lycett of the University of Kent and colleagues used information on the presence or absence of tools at chimpanzee study sites across the width of Africa to compare the distribution of tool use to the genetic data on the chimpanzee gene pools. They concluded that while regional chimpanzee populations are genetically distinguishable, their cultures are not. They occur in a mosaic that does not correspond to the genetic structure of chimpanzees across Africa. Lycett and his colleagues concluded that genetics do not play a role in population-wide chimpanzee social traditions.⁴⁴

Kevin Langergraber and Linda Vigilant, the latter of the Max Planck Institute for Evolutionary Anthropology, reexamined the genes-culture issue and concluded that, while culture certainly seems to carry the day, the role of genes cannot be ruled out. Langergraber and Vigilant say that Lycett and his collaborators haven't addressed the issue of genes versus culture, because they haven't correlated the pattern of distribution of each individual behavioral variant with patterns of genetic differences among chimpanzee communities. They note that trying to construct phylogenetic trees of chimpanzee cultures is fraught with problems due to migration and frequent local cultural extinctions and innovations. As we've seen, chimpanzees migrate among communities, carrying their social traditions with them, but they also carry their genes with them, and they end up distributing both. This tendency toward gene-culture correspondence doesn't mean that chimpanzee cultural behavior is genetically inherited. But it does mean that we can't rule that out without a more sophisticated understanding of the interplay of genes and behavior than we currently have.[45]

In addition to genetics, ecology plays a confounding factor in teasing out the various influences on chimpanzee behavior. As we've seen, the environment plays an enormous role in influencing chimpanzee behavior. The structure of the forest itself has an apparent influence not only on what chimpanzees eat but also on how they hunt, where they travel, and how they see their world. The only way to rule out the environmental effects is to consider each cultural tradition on a case-by-case basis. Gruber and seven coauthors did exactly this by considering cultural behavior in five well-studied chimpanzee communities. They compared the ecological context surrounding a particular behavior—fluid dipping (using a tool to collect honey)—and the tool set required to practice it in two communities at Kibale (Kanyawara and Ngogo) and three communities at Budongo (Sonso, Busingiro, and Kaniyo Pabidi). They considered three potential causes of variation in fluid-dipping behavior: genetics, individual developmental responses to local environment, and culture. Because the communities are geographically so near one another, genetic differentiation is very unlikely to account for minor differences observed in particular behaviors.[46] Kanyawara and Ngogo chimpanzees are already known to show tool use differences despite being genetically indistinguishable.

If the local environment leads chimpanzees to develop particular behaviors in response to it, we would expect communities with the most similar diets to employ the most similar tool sets for extractive foraging. Those communities that differ most in the way the fluid dip is practiced should have the most different diets overall. And if social learning rather than genes or the environment is responsible for particular foraging behaviors, then observers should be able to watch in real time as a newly innovated behavior, or one introduced by an immigrant, spreads through the community. This is what O'Malley and his collaborators had reported for ant fishing at Gombe. Lacking the opportunity to study this in real time, Gruber focused on the environmental hypothesis, hoping to support or refute it.

Gruber and colleagues studied differences in fluid dipping among the five communities by conducting a field experiment. They pre-prepared logs by placing honey in openings where it would only be accessible using tools and then placed these along popular chimpanzee foraging routes. Tools were provided near the logs in the form of sticks with leaves already stripped off. Motion-sensitive cameras captured the chimpanzees' response. Chimpanzees in the three Budongo communities all responded to the honey-trap experiment in the same way, using or fashioning the same tools to obtain the hidden honey. Tool use was also the same between Kanyawara and the Budongo sites. But the local ecologies are not the same in the three main areas, Kanywara, Ngogo, and Sonso. Although the chimpanzee diet prominently features ripe fruit in all three, the forests themselves differ in tree species composition, food availability, and fruit productivity. Differences in food availability and food intake between Sonso and either of the other communities were particularly striking. Lower food availability and quality did not, however, appear to promote tool use in those communities. Tool use patterns were unrelated to forest productivity, and ecological conditions were poor predictors of the repertoire of tool use for food extraction. Ecology and culture are best seen as complementary factors but, Gruber concludes, culture may exist independently of opportunities presented by the habitat.[47]

Meanwhile, it's clear that the nature of chimpanzee society plays a role in shaping the regional distribution of culture. Communities that are near one another are more likely to share cultural traits than those at a great distance. This is presumably due to migration patterns among communities, with migrants taking their repertoire of learned behaviors with them. Jason Kamilar of the University of Massachusetts, Amherst, and Joshua Marshack showed this empirically by analyzing the distribution of the thirty-nine cultural traits compiled by Whiten and colleagues years earlier. They showed that geography—longitude within Africa—and not the local environment best predicted the distribution of cultural traits. Communities in closest proximity had the most similar behavioral repertoires, with a clear gradient of dissimilarity as distance between two communities increased.[48] This finding did not address Langergraber and colleagues' claim that genetic differences have not yet been ruled out as the cause of geographic variation in behavior. But it's best to consider Langergraber as the devil's advocate. Based on a half century of field research, the presumption of social learning behind geographic differences among communities is really the only sensible approach to the study of such variation.

New cultural differences in chimpanzee behavior across Africa are being uncovered every year. Some are subtle details of tool use or body language that appear patently cultural. Others involve aspects of behavior in which the line between learned behavior and response to local habitat isn't clear. When chimpanzees prefer to catch baby monkeys rather than adults in one forest but prefer adult monkeys in another, is that due to the constraints of the forest structure lending themselves to one practice or the other? Or is it perhaps due to techniques that have been practiced by male hunters simply because older males had always hunted that way? We are still a long way from understanding exactly why some traditions become entrenched for generations, while others wink out of existence quickly or fail to take hold in the first place.

CHAPTER 9

Blood Is Thicker

THE WAY WE DEFINE a species has evolved over the centuries. For Plato and Aristotle, each species had an essential core. There was one most perfect blackbird, and all other blackbirds were slightly less perfect representations of the species. This logic extended to humans. Human diversity in appearance and skin color lent itself all too well to racist notions of purity and essentialism. Anthropology textbooks as recently as sixty years ago contained photo galleries of ethnic groups showing "the classic Chinese" or "the classic Scandinavian." The essentialist view of species held scientists in its grip for a very long time. Even Carolus Linnaeus, the grand describer and namer of so many of Earth's animals and plants, considered each species to occupy a cubbyhole, each of which was an entirely separate entity. His world view was blinkered by his religion. God had placed every creature on His Earth, each created separately and in its present form only a few thousand years ago. We now know that the view that species are distinct and forever separate is flawed. Species are gene pools, ever in flux in ways small and large.

When Edward Tyson dissected his chimpanzee in 1699 and noted the similarity among humans and the great apes, it was something even the most casual student of human anatomy could have used as evidence of shared ancestry.[1] In 1776, the chimpanzee received its biological name from German naturalist Friedrich Blumenbach. For reasons he didn't explain, Blumenbach gave chimpanzees the species name *troglodytes,* a reference to a mythical race of ancient cave dwellers. His countryman Lorenz Oken assigned chimpanzees to the genus *Pan* forty years later. The species had now been assigned its official cubbyhole in biological classification.[2] Charles Darwin's ally Thomas Huxley argued in the mid-nineteenth century that the apes should be placed in the suborder Anthropoidea along with humans, breaking with religious doctrine that would have kept us forever separate.

Darwin and Huxley both noted that chimpanzees and gorillas possessed similar anatomy, indicating a common ancestor with one another, and also with humans. Museum cabinets began to fill with skeletons and taxidermied specimens of apes. The first bonobo known to science was shot in the former Belgian Congo in 1927. For decades they were called pygmy chimpanzees, even though they're not really smaller, just a bit more slender in build. Using comparative anatomy as the only evidence, there was debate for a time about the distinctiveness of *Pan troglodytes* as a species.

Many scientists today have built their careers on comparisons of ape and human anatomy. The search for fossil chimpanzees has been more elusive than the search for human fossils, but there is evidence that the ancestors of modern chimps occupied the East African Rift valley during the same time period that the earliest members of the genus *Homo* lived there. Sally McBrearty of the University of Connecticut and Nina Jablonski of Pennsylvania State University reported on the first fossil apes clearly identifiable as nearly modern chimpanzees.[3]

The rise of modern evolutionary dating techniques in the 1970s changed our perspective of the human family tree in profound ways. The earliest biochemical revelation was made when anthropologist Vincent Sarich and the late biochemist Allan Wilson of the University of California, Berkeley, devised a way to use the immunological kinship of all animal species to ascertain the distance that separates each species in evolutionary time. Sarich and Wilson knew that blood-transported proteins were so similar among humans, chimpanzees, and gorillas that the degree of separation must be very slight. They injected a blood serum protein found in humans, apes, and other primates into rabbits and measured the strength of the antibody reaction produced. The closer the kinship was, the weaker the reaction. Sarich and Wilson were able to pinpoint the date at which early humans diverged from our common ancestor with modern chimpanzees.[4]

Devised less than twenty years after the discovery of DNA, this immunological clock was a big deal. But more advances in dating the human-ape divergence were just around the corner. In the early days of molecular genetics, it was realized that neutral molecular changes that accumulated over time amounted to a clock that could be calibrated against well-established events in the fossil record. So long as the molecules in question changed at a known and constant rate, this molecular clock could provide a precise measure of the date when two species had diverged from one another. Not all genetic systems are useful in this regard; natural selection and other forces can create confusing and inconsistent results.

The results of the molecular clock analysis had profound implications for estimates of the timing of the human-ape divergence. Sarich and Wilson's immunological results, combined with the emerging molecular evidence, pointed to a divergence between the ancestors of humans and African apes only five million years ago. At the time, scientists who studied the fossil record were convinced the divergence occurred far earlier, perhaps as much as twenty million years ago. The consensus fossil date had been reached only after decades of searching for and excavating fossils, followed by years of analysis of the specimens back in the laboratory. Sarich and Wilson needed only a short time—today the test could be performed over a lunch hour—to reach their conclusions and stake their claim. Sarich dismissed fossil candidates for the earliest human by saying that if the fossil lived earlier than five million years ago, it could not, by definition, have been a direct human ancestor, no matter what it looked like.

It was eventually shown that Sarich and Wilson's divergence times were only very slightly off; they had essentially nailed the account of human and chimpanzee evolution. In 1984, Charles Sibley and Jon Ahlquist of Yale University established a more accurate and comprehensive timeline for primate divergence using a technique called DNA hybridization. DNA molecules of two different primates are mixed together. Once the mix is heated, the double-helix-shaped spirals unzip into single-stranded molecules. One can then measure the likelihood of one species' DNA attaching to another species upon cooling. The most closely similar and therefore closely related strands require more heat energy to break their bonds. This gave Sibley and Ahlquist the index they needed to build a family tree. The results showed that chimpanzees and humans were sibling species separated by about six million years of evolution.[5] We are considerably more closely related to chimps than we are to gorillas, whom we are separated from by eight to ten million years. Orangutans were found to be further out on the lineage at about fifteen million years, with gibbons closer to twenty million years ago.

Scientists continued to work toward refining these dates. By the 1990s, multiple research groups were racing to decipher the human genome. In 1997, Mary Ellen Ruvolo of Harvard University concluded that humans and chimpanzees are more closely allied to one another than either is to gorillas, despite the obvious anatomical similarity between the two apes.[6] Pascal Gagneux of the University of California, San Diego, and his colleagues produced a molecular phylogeny for all the great apes that revealed a history of lineage splitting more complex than we had previously imagined. The tripartite divergence of early hominins and ancestors of chimpanzees and bonobos was not a threeway split. Instead, the common *Pan* genus lineage split off well after the other great apes had split with *Pan* and the first hominins. The shared lineage of chimpanzees and bonobos split further into their modern forms less than a million years ago.[7] In 1998, Morris Goodman and colleagues published similar results: an evolutionary family tree of the primates based on the genetic basis of beta globin, a component of their hemoglobin molecule. Covering sixty primate species, it helped to reconcile differences between the molecular genetic evidence and the fossil record. Goodman found the last common ancestor of all primates to have lived sixty-three million years ago, astonishingly close to the fossil record estimates of sixty-five million years ago. He found the ape lineage splitting off from the monkeys twenty-five million years ago (which fits the evidence for the most ancient ape fossil, which dates to nearly twenty-five million years ago as well). Goodman and colleagues reported gorillas splitting from the human-chimpanzee cluster at eight million years ago, with humans and chimpanzees finally splitting from their common ancestor six million years ago.[8]

The 1.6 Percent

Many of us have heard that humans and chimpanzees "share 98.4 percent" of our DNA sequence. This statistic has nearly become a cliché, but how geneticists arrived at the figure and what it truly means are far from clear. Feng-Chi Chen and Wen-Hsiung Li from the University of Chicago and National Taiwan University found a greater than 98 percent overlap in the DNA sequences of humans and chimpanzees.[9] Other studies in the 2000s found between a 98.6 percent and 99.4 percent similarity. The 98.4 percent figure that is so widely noted is simply the average of at least thirty studies that produced varying estimates. These statistics, however, need to be seen in evolutionary perspective. Humans and most other mammals share about the same number of base pairs in their genomes, approximately three billion. A human has roughly the same number of genes as a mouse or dog. In most cases there are counterpart genes in each species as well. So although it sounds impressive to say that humans and chimpanzees have nearly identical DNA sequences, we also share some 80 to 90 percent of our genome with mice.

The best way to understand human uniqueness and therefore also get a handle on chimpanzee-human relatedness is to consider the functional genome. Although you may hear that we have "mapped" the genome of humans and a wide variety of other species from sea urchins to gorillas, this statement is misleading. The map is only the roughest outline, within which there is an enormous amount of work to be done to identify exactly where counterpart genes differ and where they are identical between species, and learn exactly what the unique genes or gene combinations do.

The overall similarity in the genetic blueprint between humans and chimpanzees is breathtaking, but the secrets lying in the millions of evolutionary events that separate us will someday reveal what it truly means to be human. We want to know exactly where the gene or genes lie that determine our loss of body hair, our upright posture, and our language capability. These functional parts of the genome are what matter, and they may occupy only a relative handful of sites.

Estimates of the overlap in DNA sequence between chimpanzees and humans can be done in a variety of ways, yielding varying results. The human genome contains about twenty-five thousand genes. A 1.6 percent difference would mean that only about four hundred genes, a relatively minuscule number, differ between our two genomes. On the other hand, the human genome contains about three billion total base pairs, and far less than 1 percent—an estimated thirty-five million mutated base pairs—separate chimpanzees from ourselves. These thirty-five million still represent a large number of changes in the DNA sequence, but only a handful of these are likely to have resulted in actual changes of functional consequence to the amino acid sequence. One of the earliest analyses of genomic differences between humans and chimps still stands as among the most important. Wilson and Mary Claire King, the latter of Stanford University, found that more than 99 percent of all human proteins were identical in chimpanzees.[10]

Molecular biologist Ajit Varki of the University of California, San Diego, has shown that other evolutionary forces are at work too. Glycans—sugars that attach to protein and fat molecules—are involved in cell signaling. Varki's work with coauthor Tasha Atheide on glycans, in particular sialic acid, suggests key mutations two to three million years ago that altered the brain's immune system. These were loss-of-function mutations that led to the rise of the ape brain and were later involved in the evolution of a uniquely human brain.[11]

It has also been pointed out by some molecular geneticists that only a small percentage of these mutations are likely to be under natural selection: 10 to 13 percent of the genome by one estimate. Steven Dorus of Syracuse University and his collaborators estimated that only 214 genes are involved in the building and running of the human brain and central nervous system. Of these, Dorus found twenty-four that appeared to be rapidly evolving, suggesting they were under intense selection.[12] Carlos Bustamante of Stanford University studied natural selection in protein-coding genes in the human genome. He found that the genes that make us human seem to involve, among other things, our senses of smell and hearing, our digestive system, and our lack of body hair.[13] Given that, compared to most mammals, chimpanzees and humans have very similar senses of smell and hearing, the finding suggests either that there is more to learn about how chimpanzees understand their world or that these genes may have been linked in some way to other, more fundamental changes.

The most widely publicized gene that most distinguishes humans from chimpanzees is *FOXP2*, the so-called language gene—although mutations of *FOXP2* are actually associated with a variety of cognitive problems, not just language. *FOXP2* was one of the first genes known to be involved in language acquisition, and it appears in nearly the same form in a wide range of vertebrates, reflecting more than one hundred million years of conservative retention in birds and mammals. It's also found in humans, but our version of the gene is markedly different from that in chimpanzees. This makes perfect sense given that language capabilities have clearly been under intense selection in human for the past million or more years, since sometime after our evolutionary split with the genus *Pan*.[14]

Body hair loss in humans has been studied via strongly circumstantial genomic evidence: the evolution of lice. The evolutionary split between head lice and body lice suggests that our direct ancestors lost their body hair sometime over a million years ago—perhaps much earlier than that. Alan Rogers of the University of Utah and his colleagues, using the *MC1R* skin pigment protein gene, estimated that dark skin pigment only appeared in humans around one and a quarter million years ago. Chimpanzees, under their coarse black body hair, typically have pale skin, while exposed skin is darker. Their skin is protected from the sun's harmful ultraviolet rays by hair, and also by being under tree cover most of the time. As soon as humans evolved a loss of body hair, skin cancer became a health threat. At that point natural selection would have exerted strong pressure for more protective melanin in the skin.[15]

Among the genes that are functionally unique to humans, Caleb Finch and I focused on apolipoprotein E. *APOE* is involved in fat metabolism and facilitates the uptake of cholesterol and lipids throughout the body. The alleles of *APOE* are implicated in a wide range of diseases, including cardiovascular disease and Alzheimer's disease. There are three alleles of the gene—*APOE2, APOE3,* and *APOE4*—of which *APOE3* is the most common in humans. *APOE4* is a major indicator of future Alzheimer's risk and, to a lesser extent, cardiovascular disease. *APOE3* apparently appeared in the human genome in the past several hundred thousand years and may have originated from an ancestral *APOE4*-like gene. In other words, *APOE4* appears to have been the ancestral allele in all primates, whereas *APOE3* appeared only in the hominin lineage. Because of its role in mediating fat and cholesterol in the body, and its appearance soon after the human-ape evolutionary split, Finch and I hypothesized that *APOE3* is an excellent candidate for an early human adaptation that would have allowed for a meatier diet without the disease risks that would otherwise accompany it. The long human life span—decades longer than that of chimpanzees—might be tied to a gene that mutated in our ancestors to allow them to eat a higher-quality diet while avoiding associated diseases. Instead of shortening our life spans—as has been shown to occur in rodents fed a high-calorie or high-fat diet—it appears to have done the opposite. Finch and I argued that "meat-adapted" genes likely underwent mutations that made this paradox possible.[16]

While *APOE* is a hypothetical candidate for disease-related mutation that made humans unique from the great apes, we have other disease candidates for which solid genetic evidence already exists. Human immunodeficiency virus (HIV) originated in central Africa between 80 and 130 years ago. We've studied the genomes of multiple strains of HIV and now understand its evolutionary history fairly well. HIV underwent an evolutionary split from its ancestor, simian immunodeficiency virus (SIV), which occurred millions of years ago in central or West Africa in several monkey species. We now know of at least forty strains of SIV, two of which evolved long ago into HIV-1 and HIV-2, and very few chimpanzee populations appear to be free of the virus. The SIV strain that became HIV-1 is in fact named SIVcpz to indicate its chimpanzee ancestry. Most likely, chimpanzees occasionally hunted and ate the monkeys, and humans occasionally hunted and ate chimpanzees, creating a chain of transmission. The discovery of SIV led to field tests via urine or fecal samples of the prevalence of the virus in wild chimpanzee populations. The Gombe chimpanzees were tested; 9 to 18 percent tested positive. It didn't appear that SIV was a lethal condition, however. Unlike AIDS, which inexorably progresses in an HIV-positive person, SIV seemed to be harbored by chimpanzees without obvious harm.[17]

For a decade following the discovery of SIV, it was assumed that chimpanzees were only carriers of harmless SIV. But further research at Gombe showed otherwise. Brandon Keele and Beatrice Hahn of the University of Alabama, Birmingham, and their collaborators showed that SIV-positive Gombe chimpanzees were ten to sixteen times more likely to die than uninfected chimps. SIV-positive females had higher rates of infant mortality. As in people stricken with AIDS, T-cell counts were depleted in SIV-positive chimpanzees. Some infected chimps nonetheless lived long lives, with no evidence of an inevitable progression to AIDS-like ailments in infected individuals. Whether the overall life expectancy of SIV-positive chimpanzees differs from that of unmedicated HIV-positive human victims is not yet unknown. Molecular studies of the virus suggested that SIV has been infecting chimpanzees for several thousand years, far longer than HIV has affected humans.[18] It may be that chimpanzees evolved some degree of immunity to SIV over the hundreds of generations they've been infected with it.

When we compare humans and chimpanzees, we have to bear in mind that since our divergence from a common ancestor, each lineage has undergone millions of years of evolution. Two recent studies, one by Priya Moorjani, Molly Przeworski, and colleagues of Columbia University and the other by Kevin Langergraber and his coauthors, estimate a divergence at least eight million years ago in the human-chimpanzee lineage.[19] That's at least sixteen million years of combined accumulations of new mutations. However small the genetic changes may have been, the anatomical changes have been dramatic. Wilson and King showed that differences in gene expression, rather than gene structure, were likely keys to making us human. If humans and chimpanzees were two species of frogs or other animals, we might classify ourselves in separate suborders based solely on anatomy.[20] In other words, our genomes and our bodies have followed quite different evolutionary paths.

Chimpanzee as a Species

The question of how to distinguish humans and chimpanzees in evolutionary terms is attracting new avenues of research and producing some answers. But the question of how to define a chimpanzee is not straightforward. Today we recognize that species are like clouds with fuzzy borders that merge into other, closely related species. They are gene pools in a state of some flux. This recognition of the fluidity of species has had profound implications for understanding the process of speciation. Biologists argue over which of many concepts of species to employ. Populations that have been separated for eons so that no interbreeding has taken place—such as on oceanic islands or in patches of forest separated by rivers or human development—are often considered species. This is despite the fact that a male from one of the long-isolated populations and a female from the other might breed readily and even produce fertile offspring if placed together in captivity. The application of evolutionary principles to what we see in nature is not always a smooth and easy fit.

Chimpanzees share a genus, *Pan,* with the closely related bonobo. Molecular genetic evidence indicates the two shared a common ancestor

The average degree of genetic relatedness in several chimpanzee communities.

Community	Size	Percent Paternal Half Siblings
Gombe Kasekela	45	11.0
Mahale M group	60	15.6
Taï North	38	18.6
Ngogo	150	5.1

* Data source: Inoue et al. 2008

a bit less than a million years ago. Around the time that *Homo erectus* was expanding its range out of Africa and into Asia, an ape in central Africa split into two or more nascent species. Two populations were isolated from one another, and mutations accumulated until the two gene pools were genetically and anatomically distinct. Behavioral primatologists would point to some fairly stark differences in sexual and social behavior as distinguishing hallmarks of each species as well. Kay Prüfer of the Max Planck Institute for Evolutionary Anthropology and colleagues showed that 3 percent of the modern human genome is more similar to either the chimpanzee or the bonobo genome than either is to each other.[21] This speaks to the close sibling species status that we have with both apes.

Chimpanzees are currently divided into four subspecies, although a recent study by Javier Prado-Martinez of the Sanger Institute and colleagues shows some to be far more divergent than others and suggests the current classification scheme may need revising.[22] How we define a subspecies has been fraught with as much rancor as any other topic in evolutionary biology over the past four decades. Depending on whom you ask, a subspecies is a distinct population that appears to be on its way toward a full speciation event, an incipient species in the making; it is an anatomically or genetically distinct unit that merits its own name simply because of its distinctiveness; or it is an arbitrary and meaningless splitting of a species into smaller units. You can find all these positions (and more) argued for in the pages of biological journals.

Sometimes, biologists want to split a species into more subspecies so badly that they ignore evolutionary reality. In a paper about mountain gorilla classification, anatomist Esteban Sarmiento and his colleagues tried to make the argument that the mountain gorillas of the Virunga volcanoes should be reclassified apart from those in nearby Bwindi Impenetrable

National Park, Uganda. They cited a list of modest anatomical differences, few of which held up to statistical scrutiny. A study of the two populations by Karen Garner and Oliver Ryder of the San Diego Zoo confirmed that the two populations are genetically indistinguishable.[23] The once-vast expanse of forest in the region has been fragmented in the past few centuries, creating a patchwork in which the two gorilla populations looked more separate than they actually were. With only eight hundred of the apes in the combined forests, dividing mountain gorillas into two subspecies of four hundred each would have had profound implications for the conservation management of the species.

None of the currently recognized subspecies of chimpanzees is critically small, but that will likely not be true a century from now. Gaining a full understanding of the nature and scope of chimpanzee genetic variation is therefore vitally important. *Pan troglodytes* consists of *P. t. troglodytes,* the nominate subspecies found across a wide area of central African rain forest. *P. t. verus* is the western subspecies found in Taï National Park and surrounding forests. *P. t. schweinfurthii,* the eastern chimpanzee, is the subspecies that has been the most studied and has the largest distribution, in Gombe, Mahale, Budongo, and elsewhere. A fourth subspecies, *P. t. ellioti*—the Nigerian-Cameroonian chimpanzee—has been named in recent years on the basis of mitochondrial DNA evidence analyzed by Katherine Gonder and colleagues at City University of New York.[24] The boundaries of these categories are not entirely agreed on. Anne Fischer and her collaborators at the Max Planck Institute for Evolutionary Anthropology showed that the degree of genetic differentiation among the chimpanzee subspecies, based on nuclear nonrepetitive DNA sequences, was about the same as among human ethnic groups.[25] Scientists don't think it makes sense to pigeonhole different human ethnic groups as subspecies. Fischer and colleagues argue that it doesn't make sense to do so for chimpanzees either.

Most species that occur over a wide geographic area exhibit both anatomical and genetic variation. Populations vary incrementally and often subtly. The greater the distance between two populations, the more genetically differentiated they are. Studies of chimpanzee subspecies have typically used genetic samples that were unfortunately taken from unidentified locations within the huge chimpanzee range. Thomas Fünfstück and colleagues, also of the Max Planck Institute, used microsatellite genotypes from known locations and argued that chimpanzee variation is best understood as a gradual cline rather than a series of traditional subspecies, each slotted into its own box. This is also how we view the genetic variation of modern people. Chimpanzee populations found along the equator from East and central Africa clumped together more closely than the named taxonomic boxes did. This is surprising, since there is a large area of arid country in West Africa—the Dahomey Gap—that is believed to be the result of cooling and drying climate trends and the retreat of rain forests in the past million or so years. This climate change was thought to have isolated central and western chimpanzees and led to their genetic differentiation. But Fischer and colleagues argued that the separation of the two populations may be of more recent origin than that, too short a time to account for much evolutionary divergence between them.[26]

Meanwhile, genetic analyses are even rewriting the time-honored story of how chimpanzees and bonobos came to be. It has been assumed for decades that chimpanzees and bonobos diverged from a common ancestor between one and two million years ago when the vast Congo River changed course. This would have created an impassable geographic barrier, cutting two ape populations off from one another. In the absence of gene flow between the populations, they evolved on their own paths, producing the two modern-day species of the genus *Pan*. But Hiroyuki Takemoto and colleagues of the Primate Research Institute of Kyoto University argue for a different scenario. Based on recent evidence that the Congo River is far older than we had previously thought, combined with the molecular data on the ancestry of bonobos, the authors argue that the ancestral apes that became bonobos crossed the Congo River at shallow points during a period of exceptionally low water. After the river level rose again, that small population was trapped on the left bank and over time diverged from the apes on the river's other bank.[27]

A better understanding of the genetic structure of chimpanzees can help us make and implement conservation strategies. It may also help us to dissect aspects of social behavior and demography. Studies of relatedness among males using DNA analyses have informed us enormously about the political fabric of chimpanzee society. Given that communities are composed of a core of males that were born there and do not migrate, we might assume a high degree of kinship among the males. But this appears not to be necessarily the case. A 1990s study of the Kanyawara chimpanzees by Anthony Goldberg and Richard Wrangham of Harvard University showed that the average degree of relatedness among males was lower than expected.[28] More recent results have been equivocal. Linda Vigilant and her colleagues showed that relatedness among male Taï chimpanzees was not significantly higher than among females. At Taï, the alpha fathered more than half the offspring, and in eras when there were many males in the community, the alpha's ability to dominant matings decreased. This was a surprising result given that there is practically no gene flow from males from other local communities. Indeed, the authors found one case of likely paternity from outside the community in dozens of genetic samples. They proposed that we reassess our belief that male cooperation in chimpanzees stems from shared genes. Instead, males of a given community have a strong incentive to bond in territorial defense regardless of kinship. Bonds within the community between males and females may also trump those among males, depending on the circumstance.[29]

The same effect appears to be true for other chimpanzee populations, with some twists. Eiji Inoue of Kyoto University and his collaborators showed that the average relatedness among males in Mahale Mountains National Park was higher than that among females. This may be because a single alpha male (Ntologi) held his tenure for so many years that his genes ended up in a large number of his sons, who remained in the community while his daughters emigrated. It appears that the average genetic relatedness among males in chimpanzee communities, despite male philopatry, rarely reaches the level of half siblings. It is, however, usually at least marginally greater among males than among females. This level of relatedness has some effect on male cooperation, since related males spend a great deal of time together. Male cooperation that is not directly kinship based nevertheless has the same appearance and effect. In chimpanzee society, having the same father but different mothers leads to a level of male cooperation that is effectively kin selected. In the Inoue study of paternity at Mahale, the alpha male fathered nearly half of all the offspring. This is similar to the result at Taï (50 percent) and Gombe (35 percent). Among the half of offspring that don't migrate (that is, the males), we expect a degree of relatedness approaching 25 percent. In fact, the degree of relatedness among males is slightly less in all communities studied except

Ngogo, where it is only 5 percent, due in all likelihood to the large number of males in the community who are fathering offspring.[30]

Langergraber and his collaborators used the Y chromosome to study community structure and life span. Since the Y chromosome is inherited from father to son in each generation, it's ideal for studying species in which related males tend to live together. It has been used widely in research on human societies but rarely to study nonhuman primates. Chimpanzee communities are, as we've seen, highly patrilocal—a cohort of males are born, live their lives, and die in the same place, while females migrate. Langergraber and colleagues estimated times to the last common ancestor of the males of one chimpanzee community. They genotyped nearly three hundred chimps from five communities at thirteen gene locations across Africa using DNA extracted from fecal samples. By comparing alleles within each gene locus between potential fathers and sons, Langergraber and colleagues ascertained that it had been from a few hundred to two thousand years since the males of each sampled community had last shared a common ancestor.[31]

Put another way, some chimpanzee communities have stayed together for as many as 130 generations, a chimpanzee generation being about sixteen years. This seems like an extraordinary degree of social stability, given that we have seen the near annihilation of at least two communities (at Gombe and Mahale) in the span of fifty-five years of observation. Because chimpanzee communities are male-centered, the size and stability of male cohorts may strengthen their ability to control territory and females. There is clearly a breaking point though, as community fission and ensuing intergroup conflict in the 1970s at Gombe showed.

Ultimately, many researchers studying chimpanzee genomics do so with an eye toward understanding patterns of human evolution. Genomic comparisons may uncover evolutionary forces that were at work in the emergence of the earliest hominins. Researchers working with fossils and molecular genetics agree on the roughly six-million-year divergence time between our lineages. Nick Patterson and his colleagues at Harvard University and the Massachusetts Institute of Technology argue that the real divergence may have been substantially more recent, with frequent hybridizing between our ancestors perhaps long after the split.[32]

Genomic tools for analyzing patterns of human evolution and migration become more powerful every year. But they are equally powerful for understanding the nature of chimpanzee society, and for their role in the conservation genetics of endangered species. The use of molecular genetics has revolutionized the study of rare animals. Even in forests where hunting is intense and the animals are too shy to be observed, we can learn a great deal about them by collecting genetic samples. Maureen McCarthy and colleagues used genetic techniques to study the social structure of chimpanzees in fragmented forests in Uganda. The encroachment of farms and villages into the forest has created a patchwork in which we presume chimps are the losers. In a recent paper, McCarthy presented her evidence of the size of the chimpanzee population in her study area based on genotyping at a number of microsatellite loci. She identified at least 182 chimpanzees and a population size in her area of about 256 animals. The clumped distribution of the genotypes suggested at least nine communities with eight to thirty-three chimpanzees each. This population estimate was more than triple that of earlier censuses based on counting chimpanzee nests in small areas and then extrapolating to the whole region. If McCarthy's results hold true more widely, they suggest that (1) there may be more chimpanzees in human-settled areas than we thought, and (2) genetic techniques using collected poop samples may be a far better way to census rare animals than the more time-honored census methods.[33]

CHAPTER 10

Ape into Human

FROM THE MOMENT that apes reached our consciousness, scientists have wondered what chimpanzees can tell us about what it means to be human. The earliest observers studied our similarities, while zoogoers noted the ways in which chimpanzee behavior resembled our own. Charles Darwin wrote at length about the great ape–human connection, his frame of reference being orangutans in the London Zoo.[1] Research on chimpanzees in captivity grew during the twentieth century, with Robert Yerkes and others attempting to teach language and even speech to apes. Finally, just past midcentury, Jane Goodall went to Africa to study chimpanzee behavior in the setting where it had evolved, sent there by paleoanthropologist and fossil hunter extraordinaire Louis Leakey. He saw the shores of Gombe as a reasonable analog for the ancient lakeshore that his East African fossil site Olduvai Gorge had been millions of years earlier. Goodall's project was followed by long-term studies at Mahale and Budongo, and then elsewhere. The connection between modern chimpanzees and human origins grew more tantalizing but still existed mainly in the realm of human imagination.

By the 1960s, Sherwood Washburn was perhaps the world's most influential human evolutionary biologist. He had been trained as a functional morphologist, correlating the anatomy of nonhuman primates with their behavior, with extrapolations to humans. As a young scholar on the Asiatic Primate Expedition in 1937, he spent weeks in the rain forests of Thailand. There he dissected gibbons and monkeys to examine the workings of their joints and muscles, only hours after watching them alive in the forest canopy. As a faculty member at the University of Chicago and later at the University of California, Berkeley, he conceived the idea that what was then called physical anthropology (now known as biological anthropology or human evolutionary biology) should incorporate the study of primate behavior. Graduate students were sent to the field to undertake research on the social behavior, ecology, and functional morphology of apes and monkeys, with an eye toward reconstructing the behavior of the earliest humans.

Washburn saw the chimpanzee as an excellent analog for early humans. In the 1960s, it seemed clear that our earliest ancestors had been savanna-living, big-game-hunting, bipedal apes. Chimpanzees do not actually lend much support to the notion of early hominins as hunting, upright-walking apes. They are not big game hunters, nor are they bipedal. Nevertheless, Washburn pushed his *troglodytian* model, seeing chimpanzees as anatomically appropriate proxies for early humans. He suggested that knuckle walking might have characterized our earliest hominin ancestors as a transitional stage to bipedal walking.[2] The emerging field of biochemical approaches to human evolution provided evidence for a very recent divergence between the hominin and ape lineages, supporting Washburn's theory. Few modern scholars accept Washburn's idea of a knuckle-walking stage in early humans. The earliest apelike hominins had likely begun their break with ape biology with upright posture as a key adaptation for exploiting new habitats and new resources.

While these early chimpanzee models seem naïve to us today, the debate continues. There are at least two types of theories that have been built with information from chimpanzees. The first is to assume that the earliest hominins were quite chimp-like and reverse engineer the common ancestor from them. This was certainly the early approach. The other is to use the many elements of chimpanzee biology and triangulate the likeliest behaviors in early humans. What does chimpanzee tool use suggest about the origins of human tool use, and what do the patterns of chimpanzee meat eating suggest about the advent of meat eating in hominins? This is the more modern approach, although it too is fraught with methodological problems, not to mention a lack of hard data from fossils. Instead of seeking to paint a portrait of early humans, we're better off establishing a reality check of sorts. We bracket the range of behaviors we might expect to find in earliest hominins with the comparable behaviors that modern apes actually engage in.

A fundamental issue is what we mean we say "early human." The last common ancestor of chimpanzees, bonobos, and humans lived about six million years ago. Are we trying to understand that last shared ancestor's behavior? Or perhaps we're trying to reconstruct the lives of hominins who lived just after the split, five to six million years ago. This was a key time period about which we knew very little until the past decade. Or maybe we can use chimpanzees to model aspects of the behavior of any early human species in the fossil record. This would include the earliest members of our own genus *Homo*. Human evolutionary biologists have tended to choose traits in chimpanzees that seemed to apply to whichever early human species they chose to fit them to. This cherry picking makes for good stories but not necessarily good science, as we'll see.

Chimpanzees and the Earliest Humans

It's important to remember that nearly every scholar who has discovered and described a human fossil has had an intellectual ax to grind. The researcher has a particular narrative that suits his or her world view, training, and ongoing research agenda. For example, the discovery and description by Donald Johanson of Arizona State University and Tim White of the University of California, Berkeley, of the famed australopithecine fossil Lucy—the first-discovered and best-known specimen of *Australopithecus afarensis*—led to a new narrative about human origins. They claimed that our ancestors were upright walkers fully committed to a terrestrial life on the savanna.[3] This fit the world view of the functional morphologists involved in the initial descriptions, in particular C. Owen Lovejoy of Kent State University. Lovejoy studies the biomechanics of bipedal gait and drew a rigid dichotomy of bipedal versus quadrupedal posture. He regarded Lucy's anatomy as bipedal.[4]

Other scholars who had the chance to study the same fossil disagreed. Randall Susman, Jack Stern, and William Jungers of Stony Brook University examined Lucy's anatomy and concluded that she possessed skeletal adaptations to both bipedal walking and tree climbing. She was a more behaviorally versatile creature than Lovejoy and company imagined her to be. Living in a habitat rich with large predators where many of the desired foods are located in trees, one would expect an apelike hominin to have retained highly adapted tree-climbing skills whether the skeleton suggested it or not.[5]

This raises an important point about interpreting behavior from fossils. Functional morphologists believe they can reconstruct how fossilized animals behaved when they were alive by analyzing the skeleton using modern primates as analogs. Where there is no modern analog—there are no mammalian bipeds other than us in the modern world—they extrapolate from diverse sources to paint a portrait of an early biped. The problem with this is the problem with functional morphology itself: there is an enormous margin of error. For all the amazing ways in which animals are adapted to their worlds, there is usually a good deal of plasticity in their biologies in order to cope with variable environments. Chimpanzees are marvelously adapted to a life in the trees, but they spend most of their waking hours on the ground. Few if any functional morphologists would interpret this correctly if they had never seen a living great ape. For all their arm length, shoulder rotation, and ability to swing from branches, chimpanzees travel almost exclusively by knuckle walking on the ground for many hours every day. Only when they need to climb a tree to reach food or a nest do they put their arboreal adaptations to work. This degree of plasticity does not show up in their skeletons. Likewise, *Australopithecus afarensis,* even if adapted to upright walking—the exact nature of that bipedal posture and gait being still much debated—would have retained substantial climbing and arboreal foraging in its behavioral repertoire. Lovejoy and others claimed that once a shift to bipedal posture and anatomy was made, it was virtually impossible for that biped to be agile in trees. This is a claim that can only be made by scholars who are not familiar with living apes, or with modern people who are adept tree climbers.

Lovejoy's perspective on human origins was unusual in its rigid view of what a strict biped Lucy might have been. He also has been a critic of the use of chimpanzees in understanding the behavior of early humans. Ken Sayers of Georgia State University and Lovejoy argued that chimpanzee-based models lead us astray in their presumptions that early hominins were very chimp-like. They claim that chimpanzees make poor models for early humans such as *Australopithecus afarensis* or the more recently discovered *Ardipithecus ramidus* because of fundamental differences in four areas: posture, meat eating, tool use, and culture. In a series of papers,

Lovejoy and his colleagues went so far as to claim that *Ardipithecus* doesn't remotely resemble the evolutionary trajectory taken by any of the modern apes.[6] Therefore, they say, using chimpanzees to understand human evolution is a fool's errand. This is a silly claim, given that without modern apes, Lovejoy would likely have been unable to grasp even the basic biology of any early hominin. The margin of error in interpretation for early humans would be as great as it is for paleontologists trying to understand the lives of dinosaurs.

No primatologist has ever claimed that early humans were just like modern apes. We simply argue that, lacking more fossils, modern great apes are one obvious prism through which to view *Australopithecus, Ardipithecus,* and kin. These fossil humans were not briefly in transition between ancient apes and humans, as White and colleagues point out. They represent an evolutionary experiment that lasted a very long time and may have given rise to our direct ancestors. But three of the four areas of evidence cited by Lovejoy have no bearing whatsoever in the argument; there is no evidence at all about whether or how *Ardipithecus* ate meat, used tools, or displayed cultural differences among its populations. It's not even indisputably clear that *Ardipithecus* is more related to hominins than to African Miocene apes. Sayers and Lovejoy argue that we should reject chimpanzee models of hunting because wolves and other carnivores also hunt. They say that because other primates use tools and display culture (the definition of culture being in perpetual debate), chimps aren't the right model species. If there is logic to their argument, it escapes most human evolutionary biologists, including me. To try to rule chimpanzees out as models requires pretending that early humans were nothing like modern apes, despite all evidence to the contrary. Lovejoy's view is ironic following a half century of hard-won knowledge of chimpanzee biology.

Let's consider how our knowledge of chimpanzee biology enhances and influences our interpretations of early humans, using the two best-known recently discovered early hominin species: *Ardipithecus ramidus* and *Homo naledi*. *Ardipithecus* is an early hominin, or perhaps prehominin, that differs from the others science has seen. They appear to represent a transition to bipedal posture and movement while retaining a suggestion of climbing limbs. They had canine teeth that were smaller than those of any modern ape but larger than those of later hominins. Lovejoy cites *Ardipithecus*'s modest sexual dimorphism in body size and canine tooth size to infer monogamy, and he speculates further to wonder whether these earliest hominins represent a transition from the chimpanzee-like sexual swellings to the concealed ovulation of human females.

There is nothing in the skeletal biology of *Ardipithecus* that suggests pair bonds, males taking care of females, or a loss of sexual swellings. This narrative of the evolution of monogamy is utterly at odds with a mountain of evidence from chimpanzees and bonobos about the evolutionary interplay of sexual dimorphism, male competition, and polygyny. Our earliest apelike ancestors were in all likelihood as far from monogamous as are all the living African apes. Monogamy is more common among primates than other mammals, but it is still quite rare. It would be simplistic to even generalize modern humans as monogamous. As for tool use, we know that stone tools were in use by around three million years ago. We have no frame of reference for making even educated guesses about when tool use arose and what form it took in earlier ancestors like *Ardipithecus*. Because tool cultures are varied and widespread among chimpanzees, they likely existed in very early hominins too. Ignoring the evidence from chimpanzees, and also observations of tool use by South American capuchin monkeys, is counterproductive. Chimpanzees also relish meat. If we want to understand what meat-eating habits may have looked like among the earliest humans, chimpanzees are a logical analog. So a chimpanzee model is the most reasonable approach to making educated guesses about culture in six-million-year-old ancestors.[7]

How Chimpanzees Help Us Understand the Past: The Case of *Homo naledi*

The most exciting human fossil discovery in recent years was *Homo naledi*, announced in 2015 by Lee Berger of the University of the Witwatersrand and his colleagues. Discovered by cave explorers, the fossils represent more than 1,500 pieces of at least fifteen individuals; they were retrieved from a chamber called the Dinaledi Chamber deep underground in the Rising Star cave system in South Africa. The *H. naledi* fossils were found in a chamber so inaccessible that the excavation was carried out by a team of spelunking paleoanthropologists. This is first truly new member of our genus *Homo* described in many years. The first analyses revealed a quite primitive lower body with a more modern skull, a mosaic of apelike and human traits.[8] The geological context of *H. naledi* is not very amenable to ascertaining the age of the fossils, though there are estimates in the works. In all likelihood the *H. naledi* fossils date to about a million years ago. Given that the fossils are clearly far more humanlike than *Ardipithecus* or any australopithecine, and the date is millions of years more recent than the earliest hominins, the study of wild chimpanzees would seem to provide little insight into the life of *H. naledi*. But there are important inferences to be drawn from great ape biology about the behavior of *H. naledi*.

Analysis of the fossils and the site itself was done by geologist Paul Dirks of James Cook University and colleagues. They believe that the far reaches of the cave in which the fossils were found was a burial chamber where *H. naledi* deposited their dead. They claim that the skeletons appear to have arrived in the chamber intact and only fell apart later as their bodies decomposed. Moreover, practically no fossils other than the hominins were found in the chamber.[9] Paleoanthropologists coming from a human mortuary frame of reference might find this entirely plausible. Early humans are known to have carried or dragged their deceased loved ones into the dark reaches of a cave as a final resting place. But to a primatologist with a great ape frame of reference, it makes little sense. *H. naledi* was a small-bodied, apelike human, much like its relative *Homo habilis*. *H. habilis* is among the most primitive members of the genus *Homo*; according to some scholars, they belong in the genus *Australopithecus*. They may have been well on the road to humanity, but they were still apelike.

Wild chimpanzees spend their days traversing a large, forested landscape. In arid regions their home range can be enormous—up to three hundred square kilometers. Even in verdant forest they will use fifty square kilometers. When chimpanzees die, whether by accident or illness, their final resting place is wherever they happen to be. Chimpanzees and other apes display an intense curiosity about death—they will stay by a dead group-mate, poking and prodding the body for hours of time before moving on. Like all primates and most animals, chimps use a portion of their home range most intensively; we call this the core area. Modern foraging people—the so-called hunter-gatherers—do the same but with a more nomadic tendency to migrate with the seasons as animal herds move. If either a great ape or a hunter-gatherer intended to transport a dead group member to a burial place, it would likely involve carrying the corpse for hours. And then, in the case of the Dinaledi Chamber of *H. naledi,* the body would have to be dragged deep into the back of the cave. It's hard to imagine a primitive, apelike hominin having a mortuary practice that involves carrying a dead body for many kilometers to a burial ground.

Other scholars are skeptical that the cave was a burial chamber. Aurore Val of the University of the Witwatersrand has challenged Dirks's and Berger's burial chamber interpretation, arguing that the authors' claims that the bodies arrived in the cave intact and have never been moved by sediments or scavengers are on shaky ground. Val argues that the difficulty of access to the chamber makes it unlikely that *H. naledi* could have carried corpses there. Far more research on the context of the deposit is needed before any conclusion is reached about the chamber's nature. The logic from chimpanzee biology is persuasive. We have no evidence of any living nonhuman creature using burial chambers. Even death areas for elephants and other animals are more legend than fact. While further studies of the chamber may support or refute Berger's and Dirks and colleagues' claims, the insight from modern apes says that the Dinaledi Chamber is very unlikely to have been a burial site. It may be that the cave was once more open to outside air than it currently is and *H. naledi* took refuge there, sometimes dying there. Much later the cave may have been closed in and buried deep by geological processes. We won't know the true meaning of the fossils or the cave without further research, but one key piece of ape biology can tell us what the cave is very likely not.[10]

No one would argue that ancient humans were essentially modern apes, any more than we would claim that chimpanzees are just downsized gorillas. Each species must be understood in its own unique context. But evidence from the biology of chimpanzees, as well as the other nonhuman primates, integrated with fossil information and paleogenomics, provides powerful tools for understanding where we came from.

What Have We Learned?

How has our understanding of chimpanzees changed over the past fifteen or twenty years? Chimpanzees have been studied in the wild since the early 1960s, so this represents only the most recent one-third of the history of research. And yet, we have learned a great deal. Some new perspectives have resulted from new analyses of ongoing long-term studies. We have, for example, new insights into why chimpanzees hunt and how their food calls are understood by other members of the community. We now understand the long-term benefits and costs of social dominance, the fate of orphans, and the natural and routine character of violence between communities. Other insights are based on brand-new observations. These are not only exciting; they suggest more discoveries still to come. They include hunting with "spears" at Fongoli, accumulations of stones under trees where they have been systematically tossed for generations, new observations of violence within communities, and new complex tool sets.

As new discoveries are made, they allow us to see patterns in observations that formerly didn't seem very connected. Andrew Whiten's compilation of chimpanzee culture in the late 1990s produced thirty-nine behaviors that are clearly cultural and occur systematically.[11] In a more recent paper, Whiten collaborated with archaeologists Nicolas Toth and Kathy Schick to compare and integrate information on stone tool use from primates and the fossil record. They examined the ways in which tool cultures of chimpanzees inform us about those early members of the genus *Homo,* and vice versa. Whiten and colleagues point to the ever-expanding catalog of chimpanzee percussive stone tool uses that indicate the potential power of innovation in both apes and primitive humans.[12] Cultural diversity among chimpanzee populations certainly suggests the same for the early members of our own genus. We have Oldowan tools—very primitive stones modified on only one edge but useful as efficient cutters and choppers. In all likelihood these tools varied a bit from one hominin site to another, the archaeological signature of which is too subtle and the sample too limited to make those assessments at present.

Michael Haslam of the University of Oxford recently weighed in on the same question, arguing that the tool use of modern chimpanzees, and the stone tool dexterity of Kanzi, suggests an early hominin transition from plant-made tools—termite probes and the like—to stone tools. He believes this started in West Africa during the Pleistocene. We've seen that cultures occur as a mosaic across Africa. Finally, experimental studies of the development and transmission of stone tool use by both chimpanzees and bonobos point to the role of innovation and social learning in fossil humans. The authors used the mapped distribution of stone tools among modern chimpanzees to infer aspects of geographic variation in similar hominin tools two or more million years ago.[13]

We can also look at what chimpanzees eat to infer aspects of the early human diet. The emergence of meat as a key part of the early human diet has been a subject near and dear to my own work for many years, but meat is not the only kind of animal protein available. William McGrew has been vocal in advocating the importance of insects as a valuable source of nutrients for both living apes and ancient humans. He points out that modern foraging and horticultural people in many societies relish insects. There is circumstantial evidence for insect eating in fossil humans too.[14] Francesco d'Errico of the University of Bordeaux and Lucinda Blackwell of the University of the Witwatersrand examined microscopic patterns of wear and tear on fossilized tools made of bone found at several sites in South Africa. They claimed that the pattern of wear on the bone tools indicates the bones were used to open earthen termite mounds.[15] No modern chimpanzee population actually does this, although chimpanzees in Goualougo in the Republic of Congo use branches as termite-mound-perforating tools.[16] The bone tools might have had other digging purposes, but their wear pattern is consistent with damage from breaking open hard earth. Without the benefit of watching wild chimpanzees termite fish, there would have been no reason to suspect the purpose of such a tool.

It takes years to develop a chimpanzee study to a point at which exciting new behavior may be seen. Long-term funding and logistical support for the project must be maintained until exciting results emerge. All long-term chimpanzee studies in which the apes were approachable enough to see subtle details of behavior have been conducted in forest environments. But at Fongoli in Senegal, Jill Pruetz and Paco Bertolani, the latter of Cambridge University, study chimps in a hot, arid, savanna landscape where they engage in behaviors we don't usually associate with chimpanzees. We've already seen that Fongoli chimpanzees are unique in using sticks to extract bush babies from tree cavities. They differ in other notable ways from chimpanzees elsewhere. During the long hot season, when temperatures reach over forty degrees Celsius and shade trees are in short supply, Fongoli chimpanzees forage at night and use caves as cool shelters. Even more evocative of early human behavior is their use of water. Most apes shun water except for drinking, but Fongoli chimpanzees soak in waterholes during the hottest months.[17]

Fongoli chimpanzee behavior may not resemble that of early humans any more than the behavior of other modern chimpanzees does, but their adaptation to an environment in which we know early hominins lived expands the likely range of behaviors we should expect in early humans. Caves, water, and regular nocturnal travel are new additions to the repertoire we've come to expect in wild chimpanzees. The Fongoli chimps also travel in larger parties, as a percentage of the overall community size, than do any other chimpanzee population. It would no longer be surprising to discover this in other chimpanzee populations in similar habitats. Speculations about the same behaviors in primitive hominins are given credence by these discoveries. When paleoanthropologists try to reconstruct early hominin behavior, they turn to environmental factors they can measure. They do this to generate realistic scenarios of how human evolution may have been guided by factors such as food, predators, seasonality, and anything else they can account for. Observations like those of the Fongoli chimps provide exactly the sort of information that one might predict from knowing what the habitat is like, if only we had a time machine.[18]

Finally, we have cases in which something new and exciting is observed in the world of chimpanzees, but we are left only with speculation to understand how it might apply to our origins. Stone throwing and stone accumulations in tree trunks are perfect examples. These are truly fascinating behaviors without obvious purposes.[19] If a scientist discovered such stone deposits in a forest where no chimpanzees were known to exist, she would probably ascribe them to the actions of modern or ancient people. Why chimpanzees would direct such energy at seemingly random trees is unknown. There is very likely a function—perhaps communication or habitat marking—that has yet to be uncovered. The authors of the paper that presents this discovery conclude with a few sentences about the meaning of apparently ritual behavior and its use in interpreting early human archaeological sites. From that cautious statement, the media concocted stories of sacred stone sites and proto-religious behavior. As I said earlier, this is nonsense. If further research fails to turn up any rational explanation for the behavior, then perhaps we are allowed to wonder what it means in philosophical rather than pragmatic, scientific terms. Until then, such speculations are just distractions.

Does Research on Chimpanzees Help Protect Them?

The past decade and a half of chimpanzee field research has greatly expanded our insight into the species. It has also, as we saw earlier, added to the lessons we can learn from chimpanzee biology about human origins and evolution. But how much has it really moved the big-picture research needle, and what big questions still remain? Chimpanzees are one of the most thoroughly studied of all nonhuman primates. That said, there is still a world of information about them that we long to know. The answers to those remaining questions will be long in coming, because in many cases the questions can't be addressed until we have generations more of data and observations. Given the rapidly declining populations of apes and their habitat, chimpanzees may not have that much time left.

In my 2012 book *Planet without Apes,* I addressed the many threats facing great apes today. Habitat destruction through deforestation in favor of farms and villages has a devastating impact, and its pace only increases. The meat of African apes is relished by people in West and central Africa, and as roads are built deeper and deeper into the remaining expansive tracts of forest, apes are among the first victims. This is what conservationists call the bushmeat crisis. Timber company workers and miners are given guns and told to find their own meals, and those meals are often wild primates. The value of ape meat is so high that an illegal export market exists, both within Africa and anywhere the African diaspora has gone, from Paris and Brussels to New York. Diseases like Ebola and anthrax have killed more gorillas and chimpanzees than they have human victims. Mining for valuable ores in great ape habitat, and the miners' hunting of the apes in those forests, has decimated eastern lowland gorillas. The list of threats to chimpanzees and the other great apes goes on and on.[20]

There is much fascinating research on wild chimpanzees that has little direct value to conservation. Some conservationists wag a finger at primatologists who conduct research on animals for years without becoming directly involved in their conservation. Many eventually do, but this criticism misses a crucial point. The key value of great ape field research to conservation has always been the presence of researchers on the ground. Many of the premier protected natural areas where great apes live today received conservation attention only because of the long-term research efforts of primatologists. Dian Fossey lived and worked in the Virunga volcanoes long before the area was designated a critical habitat for the last remaining mountain gorillas. The Mahale Mountains were the site of the second long-term chimpanzee field study. That work continues today, and Mahale Mountains National Park also protects a rich piece of biodiversity along the shore of Lake Tanganyika. The same is true for the Budongo Forest Reserve, Taï National Park, and numerous other globally important natural areas.

Ironically, the main benefit of chimpanzee field research to the chimpanzees themselves is the awareness it raises of their presence, and of the need to protect them. In the long run, this takes precedence over any of the findings the research produces. We want to know how much space chimpanzee communities need and what food species diversity is optimal for them. We can learn amazing things about tool use, cultural variation, and the like. But the most important aspect of the research is the framework it provides for a future conservation project. When the research findings are announced, they make it known to the world that a new chimpanzee culture exists that deserves protection. This information can be used to approach the host government to recommend that a protected area be established. The eventual economic benefits of ecotourism can be extolled, as well as the value of saving forests for posterity as part of a nation's heritage. Conservation groups and government officials can apply for funding from international aid agencies and conservation organizations to begin the long process of providing protection not only on paper—which in many developing countries is meaningless—but also in reality.

Establishing a protected site in which chimpanzees live has conservation implications far beyond those that pertain to the apes. Since chimpanzees use large areas of forest, protecting them means protections for all the plants and animals they share their habitat with. They become flagship species for the preservation of the entire forest ecosystem. Every living thing, from apex predators to rare plants, receives sanctuary as a result.

The research project has likely involved training a staff of local field assistants and other local people who are essential to making the project run, from porters to drivers. These people are more than employees. They become ambassadors between the researchers, who are nearly always expatriates from Western countries, and the residents of local villages.

Local people have learned to view the arrival of foreigners on their land with a degree of circumspection, and occasionally hostility. They've heard too many stories in which foreigners with cameras and binoculars ask for permission to survey their forest, and later the government decides to convert their land into sanctuaries. Sometimes this has involved expelling residents from lands they have occupied for generations without fair remuneration.

When I arrived with a small team to begin a project on chimpanzees and mountain gorillas in Bwindi Impenetrable National Park in the mid-1990s, a thriving ecotourism program already existed at the main camp at Buhoma. We established a camp in a more remote corner of the park, Nkuringo. As we hiked into our future camp for the first time, the word had already reached Nkuringo that we were coming. A line of applicants greeted me. They knew we'd need assistants to help with field research: a cook, a camp keeper, and sundry other helpers. I hired a field assistant whose uncles were the farmers who leased a plot of land adjacent to the national park on which we built our camp. He was a wonderful right-hand man during the years of the project. Two decades later, he has held local political office and continues to spread the word about great ape conservation to communities in the area. The trails that we cut and marked through the forest in the 1990s are now nature trails on which park rangers take ecotourists on scenic hikes.

Many great ape researchers become directly involved in on-the-ground conservation efforts. This often involves community outreach programs in villages surrounding the forest. It's shocking to see how many children in small villages just outside a forest with chimpanzees have never really seen a chimpanzee up close. The mission of Goodall and of many other field scientists has shifted from data collection to consciousness raising because of the moral imperative of saving the animals they've built their careers on. The more appreciation that local people in Africa have for the wild apes they share their world with, the more likely we are to achieve a lasting success in the battle to save them.

In 2012 I wrote that by the start of the twenty-second century, the great expanses of tropical forests in Africa would be gone, replaced by a patchwork quilt of towns, agriculture, cities, and remaining stands of forest. After that line was published, I was criticized by some readers for taking a cynical view of the future of the natural world. Now the twenty-second century is only some eighty years—one human lifetime—away. I see no reason to think that human development in Africa will spare more than a fraction of the land needed by wildlife populations of the present day. In fact, even this will only be possible if poaching is brought under more control in the future than in the past and present. Some primate species will undoubtedly be lost to extinction—there are just too many critically rare monkeys and lower primates. Opportunities for ape-based ecotourism may emerge in some new areas, but tourism is no panacea. It's equally likely that the cycle of political instability and war in Africa driven by poverty and abetted by dysfunctional governments will rob apes of forests, including national parks, in which they currently live. It's highly likely that by the next century, chimpanzees will barely exist outside protected areas and will be under siege even there. Everything we do today is to ensure that we achieve the bottom line: that they do not slide into extinction.

While this might sound like a bleak prognosis, it's very much what a conservationist of an earlier era might have said about North America. In 1800, the grizzly bear ranged across the Great Plains and westward to the Pacific Ocean, occupying all habitats except deserts and mountain tops. Wolf packs roamed the eastern woodlands, having been extirpated by people only from New England. Bison herds, which we now associate with Yellowstone and a few other refuges, were seen east of the Mississippi River. Today, all these magnificent animals hang on only in a few protected areas, mostly in rugged mountain ranges that were not useful for ranching or farming to nineteenth-century settlers. Passenger pigeons were the most abundant vertebrate animal in the Western Hemisphere in the late 1700s. By the late 1800s they were all but gone, their inconceivably vast flocks reduced to a trickle of doomed survivors. Had you prophesied in 1800 a twenty-first century in which all these species were either gone or reduced to a tiny fraction of their former abundance, you would have been laughed at. Any early conservation-minded listener would have regarded our century as unconscionably bleak. But nearly all these threatened species survived. Today they're a carefully managed reminder of what we once had, and their dignity as independent nations of animals lives on, greatly diminished but protected.

All the research and all the discoveries described in this book will leave a profoundly bittersweet taste if we don't preserve wild chimpanzees in the centuries to come. As the last surviving representatives of their former greatness, they will live on to another century, in each one threatened as in the last. They will be ambassadors between our own evolutionary history as apes and our present selves, the technologically advanced humans of this century who grow more divorced from our heritage with each passing generation.

NOTES

BIBLIOGRAPHY

ACKNOWLEDGMENTS

INDEX

Notes

1. Watching Chimpanzees

1. Goodall 1971.
2. De Waal 2001.
3. The International Union for Conservation of Nature Red List's most recent assessment (Humle et al. 2016) considered wild chimpanzees "endangered." This designation was upgraded from "vulnerable" in 1996, based on the estimate that at the current rate of decline, half the entire wild population will have been lost between 1975 and 2050. The union estimates the total chimpanzee population across Africa to be between 172,700 and 299,700. But this figure is based on an earlier estimate from 2003; the total population today is almost certainly lower. The estimated populations of the four currently recognized subspecies are as follows: eastern chimpanzee (*Pan troglodytes schweinfurthii*, the subspecies found at Gombe, Mahale, Kanyawara, Ngogo, and elsewhere in Tanzania, Uganda, Rwanda, Burundi, and the eastern Democratic Republic of the Congo), 173,000–248,000; central chimpanzee (*P.t. troglodytes,* found at Goualougo and elsewhere in the Republic of Congo, Gabon, Angola, Equatorial Guinea, the Central African Republic, and Cameroon), approximately 140,000; western chimpanzee (*P.t. verus,* found at Taï, Fongoli, Bossou, and elsewhere in Ivory Coast, Burkina Faso, Ghana, Mali, Liberia, Guinea, Sierra Leone, and Senegal), 18,000–65,000; and Nigeria-Cameroon chimpanzee (*P.t. ellioti,* found at Gashaka-Gumti and elsewhere in Nigeria and Cameroon), 6,000–9,000.
4. Tyson ([1699] 1972) actually compared the anatomy of chimpanzees, orangutans, and ethnic "pygmy" people from central Africa.
5. Described in Kirk 2011.
6. Described in Kortlandt 1986.

7. Nishida 1968, 1983a, 1990, 1997; Nishida, Uehara and Nyundo 1979. See also McGrew 2017.
8. Boesch and Boesch-Achermann 2000.
9. McGrew 2017.

2. Fission, Fusion, and Food

1. There are nearly a thousand species of figs worldwide, making the genus *Ficus* among the largest of all plant genera. The classification of the group is undergoing much debate and revision, because most species are dioecious but some are monoecious, and botanists don't agree on the two groups' relationship to one another.
2. Wrangham 1977.
3. Emery Thompson 2013.
4. As of 2017, the community at Ngogo in Uganda numbers nearly two hundred and is by far the largest chimpanzee community ever recorded. See Watts and Mitani 2015.
5. Janmaat, Ban, and Boesch 2013.
6. Mitani, Watts, and Muller 2002; Newton-Fisher, Reynolds, and Plumptre 2000; Goodall 1986; Stanford 1998.
7. Murray et al. 2007.
8. Miller et al. 2014.
9. Watts et al. 2012a, 2012b.
10. McCarthy, Lester, and Stanford 2017.
11. Potts, Watts, and Wrangham 2011; Potts et al. 2015.
12. N'guessan, Ortmann, and Boesch 2009; Ban et al. 2016.
13. Chancellor, Rundus, and Nyandwi 2012.
14. Newton-Fisher, Reynolds, and Plumptre 2000.
15. Hashimoto, Furuichi, and Tashiro 2001; Hashimoto et al. 2003.
16. See Crook and Gartlan 1966; Clutton-Brock and Harvey 1977.
17. Wrangham 1980.
18. Female chimpanzees may belong to more than one community simultaneously and shop for genes among them. The limitations of field research make it difficult to follow such females from one community to the next if one community is not habituated to people. See Goodall 1986; Nishida 1990; Hobaiter et al. 2014; Watts and Mitani 2015.
19. Lehmann and Boesch 2004, 2008.
20. Wittiger and Boesch 2013.
21. Wakefield 2013.
22. Matsumoto-Oda et al. 1998.
23. Wittiger and Boesch 2013.
24. During our field study of chimpanzees and gorillas in Bwindi, we found that chimpanzees there slept on the ground more frequently than in other

sites, and we believed this was likely due to the recent extirpation of leopards and other large predators. But naturalists who had visited Bwindi in the 1960s had noted ground nests too, although in those earlier days leopards and lions were both present. Ground nesting by chimpanzees is likely cultural tradition and also influenced by forest structure: some forests may be more amenable to ground sleeping than others. Gorillas in Bwindi also nest in trees on average about one night per week, whereas the nearby mountain gorillas in the Virunga volcanoes never nest in trees. This is likely due to the relative lack of tall trees in the Virungas. See also Stanford and Nkurunungi 2003.
25. Samson and Hunt, 2012, 2014; Hernandez-Aguilar 2009; Hernandez-Aguilar, Moore, and Stanford 2013.
26. Koops et al. 2006, 2012.
27. Samson and Hunt 2014.
28. Zamma 2013.
29. Stewart 2011.
30. Fedurek, Donnellan, and Slocombe 2014; Fedurek, Slocombe, and Zuberbühler 2015.
31. Fedurek et al. 2013.
32. Clark Arcadi, Robert, and Boesch 1998; Mitani, Hundley, and Murdoch 1999.
33. Clark Arcadi, Robert, and Boesch 1998.
34. Kalan and Boesch 2015; Kalan, Mundry, and Boesch 2015.
35. Slocombe et al. 2010.
36. Crockford et al. 2012.
37. Fedurek, Slocombe, and Zuberbühler 2015.
38. Langergraber, Mitani, and Vigilant 2009.

3. Politics Is War without Bloodshed

1. See DeVore 1965; Rowell 1974.
2. Sandel, Reddy, and Mitani 2017.
3. Sherrow 2012.
4. Nishida 1990; Nishida et al. 1992.
5. Nakamura et al. 2015.
6. Feldblum et al. 2015.
7. Machanda, Gilby, and Wrangham 2014.
8. Mitani 2009.
9. Arnold and Whiten 2001, 2003.
10. Watts 2000.
11. Nishida 1990.
12. Bygott 1979.
13. DeVore 1965, for example.

14. See, for instance, Langergraber, Mitani, and Vigilant 2007, 2009.
15. Muehlenbein and Watts 2010; Muller and Wrangham 2004b; Sapolsky 1992.
16. Hausfater 1975; Alberts, Buchan, and Altmann 2006; Strum 1982; Smuts 1985.
17. Stumpf and Boesch 2005.
18. Personal observation made in 1993.
19. Wroblewski et al. 2009.
20. Boesch et al. 2006.
21. Sugiyama et al. 1993.
22. Olshansky 2011.
23. McCarthy, Finch, and Stanford 2013. Because of the long chimpanzee life span, the long tenure of many alpha males, and the limited number of long-term study sites, our ability to conduct meaningful analyses of dominance rank and longevity was surprisingly limited. A future generation will be able to look into this effect in a more meaningful way.
24. Goodall 1968; Bygott 1979.
25. Newton-Fisher 2004.
26. Laporte and Zuberbühler 2010.
27. See Dixson 2012 for an exhaustive account of primate sexuality.
28. Fedurek et al. 2016.
29. Wingfield 1984.
30. Sapolsky 1992.
31. Muller and Lipson 2003; Muller and Wrangham 2004b.
32. Wittig et al. 2015.
33. Muehlenbein, Watts, and Whitten 2004.
34. Sapolsky 1992, 2005.
35. Muller and Wrangham 2004a, 2004b; Muller and Lipson 2003.
36. Rowell 1974.
37. Murray, Mane, and Pusey 2007.
38. Foerster et al. 2016.
39. Wrangham et al. 2016.
40. Fawcett 2000.
41. Wittig and Boesch 2003.
42. Pusey, Williams, and Goodall 1997.
43. Ibid., 829–830.
44. Boesch 1997.
45. Murray, Mane, and Pusey 2007.
46. Note that these males are all from the powerful Gombe F lineage of the Kasekela community. We don't fully understand whether there are genetic components to dominance status, such that members of one lineage have an inherited benefit, or whether being born into a powerful matriline is a benefit in and of itself. It is likely a combination of the two factors.

4. War for Peace

1. Boehm 1994, 1999.
2. Wrangham, Wilson, and Muller 2006; Wrangham 1999.
3. Wrangham 1996. In his book *Demonic Males,* Wrangham pointed out the gender disparity of people in prison for committing violent crimes, and he compared this to the much greater tendency of male chimpanzees to engage in lethal or severe aggression. He was widely misunderstood to be pointing to the inevitability of violence in our species. In fact, identifying possible biological influences on human behavior is the first step toward remedying them.
4. Power 1991, for instance.
5. In this event, related by Wrangham (1996) and others, Stella Brewer had established a rescue center inside a national park in Senegal, hoping to reintegrate several rescued chimps back into the wild. But the captive facility was attacked repeatedly by a group of wild males from the nearby forest, including a brazen night raid. In other words, the local wild community was so territorial that even captive animals in an enclosure were seen as necessary targets of violence.
6. Power 1991; Sussman and Marshack 2013.
7. M. L. Wilson et al. (2014) is by far the most comprehensive and detailed analysis of patterns of chimpanzee intercommunity violence.
8. Fawcett and Muhumuza 2000.
9. Nakamura and Itoh 2015.
10. Wallis, personal communication; Stanford, personal observation.
11. Watts 2004.
12. Nishida 1983b, 1996.
13. Watts 2004.
14. Muller, Kahlenberg, and Wrangham 2009.
15. Feldblum et al. 2014.
16. Stumpf and Boesch 2005, 2006 (Taï); Feldblum et al. 2014 (Gombe).
17. Newton-Fisher 2006.
18. Mech 1994 (wolves); Pusey and Packer 1994 (lions); Aureli et al. 2006 (spider monkeys).
19. Samuni et al. 2017.
20. Wrangham 1999.
21. M. L. Wilson et al. 2014.
22. Takahata 2015.
23. Goodall 1986; Wrangham 1999.
24. Although chimpanzees possess similar sensory abilities to our own—sight, hearing, and olfaction—they are clearly far more tuned in to their forest environment than the researchers following them. Chimpanzees will frequently travel a seemingly random path through undergrowth and

over hills, only to rendezvous with other chimps. No calls were heard by the researcher. The animals know the forest intimately from many years traveling it, and they may also detect odors wafting toward them that give away the presence of another member of the community. For all we know, they may even be able to hear footfalls approaching in the forest and identify the visitor by the sound, much as you can identify the person walking through your house by the sound of his or her footsteps on the floor.

25. Bates and Byrne 2009.
26. Watts and Mitani 2001; Watts, Mitani, and Sherrow 2002; Watts et al. 2006.
27. Boesch et al. 2008.
28. Watts and Mitani 2001.
29. Sobolewski, Brown, and Mitani 2012.
30. Watts and Mitani 2001; Watts, Mitani, and Sherrow 2002; Watts et al. 2006; Mitani, Watts, and Amsler 2010; Boesch et al. 2008.
31. Boesch et al. 2008.
32. Boesch et al. 2008.
33. Described in detail in Goodall 1986.
34. Feldblum et al. 2015.
35. Goodall 1986.
36. Nishida 1990; Takahata 2015.
37. Boesch et al. 2008.
38. Williams et al. 2004.
39. Pradhan, Pandit, and Van Schaik 2014.
40. Manson and Wrangham 1991; Wrangham 1999; Wrangham and Wilson 2004; Wrangham, Wilson, and Muller 2006.
41. M. L. Wilson, Hauser, and Wrangham 2001, 2007.
42. Watts et al. 2006.
43. Williams et al. 2004.
44. Hrdy 1977, 1979.
45. M. L. Wilson et al. 2014.
46. Murray, Wroblewski, and Pusey 2007.
47. Wallauer, personal communication.
48. Goodall 1986.
49. De Waal and van Roosmalen 1979.
50. De Waal and Aureli 1997; de Waal and Luttrell 1998.
51. Arnold and Whiten 2001, 2003.
52. Hartel 2015.
53. Wittig and Boesch 2003.

5. Sex and Reproduction

1. Watts 2007.
2. See Dixson 2012.
3. Kret and Tomonaga 2016.
4. Wallis 1982, 1992, 1995.
5. Hrdy 1979.
6. Emery and Whitten 2003.
7. Deschner et al. 2004.
8. Dixson and Mundy 1994. The occurrence of a baculum among mammalian orders is intriguing. Males of many nonhuman primate species possess them, but humans don't. Brindle and Opie (2016) argue that bacula evolved in species in which sperm competition exists, and not in those species in which male competition is among individuals via fighting or some sort of mating display. Although you might expect that the size of the baculum would therefore be inversely related to the size of the male's testes (because sperm production has been shown to be associated with large testis size in some species), the authors did not find this to be the case. Instead, promiscuously mating species were more likely to have a baculum, as were those species with well-defined mating seasons and those in which the act of sex lasts longest. All of this suggests that a baculum would be less useful in monogamous species. To the extent that chimpanzees are highly promiscuous, this makes sense. Whether humans are monogamous by nature is open to debate.
9. Dixson 2012.
10. Harcourt et al. 1981.
11. Emery Thompson 2013.
12. Townsend, Deschner, and Zuberbühler 2008, 2011.
13. Fallon et al. 2016.
14. Matsumoto-Oda 1999.
15. Watts 2007.
16. Stumpf and Boesch 2005, 2007.
17. Wallis 1995, 1997.
18. Nishida 1997.
19. Watts 2015.
20. Matsumoto-Oda et al. 2007; Matsumoto-Oda and Ihara 2011.
21. Duffy, Wrangham, and Silk 2007.
22. See Goodall 1986 for a more detailed account.
23. Nishida 1990, 1997.
24. Goodall 1986.
25. Wroblewski et al. 2009.
26. Muller, Emery Thompson, and Wrangham 2006.
27. Watts 1998.

28. Muller et al. 2007, 2011; Muller, Kahlenberg, and Wrangham 2009.
29. Feldblum et al. 2014.
30. Boesch et al. 2006.
31. Feldblum et al. 2014. Also see Emery Thompson 2014.
32. Emery Thompson, Newton-Fisher, and Reynolds 2006.
33. Langergraber, Rowney, Crawford, et al. 2014.
34. Kahlenberg et al. 2008.
35. Newton-Fisher 2006, 2014.
36. Stumpf et al. 2009.
37. Luncz and Boesch 2014.
38. Emery Thompson 2013.
39. Atsalis and Videan 2009.
40. Emery Thompson 2013.
41. Jones et al. 2010.
42. Hawkes and Smith 2010.
43. Emery Thompson et al. 2007.
44. Hawkes 2004.
45. Emery Thompson 2013.
46. Langergraber, Mitani, and Vigilant 2009.

6. Growing Up Chimpanzee

1. Ely et al. 2006.
2. Wallauer, personal communication.
3. Hinde and Milligan 2011.
4. Boesch 1997
5. Lehmann, Fickenscher, and Boesch 2007.
6. Emery Thompson et al. 2016.
7. Lonsdorf, Anderson, et al. 2014; Lonsdorf, Markham, et al. 2014. See also Lonsdorf, Eberly, and Pusey 2004.
8. Fröhlich, Wittig, and Pika 2016.
9. Plooij 1984.
10. Lonsdorf, Anderson, et al. 2014.
11. Murray et al. 2014.
12. Heintz 2013.
13. Lonsdorf 2005; Lonsdorf, Anderson, et al. 2014.
14. Goodall 1968, 1971.
15. Botero, MacDonald, and Miller 2013.
16. Kalcher-Sommersguter et al. 2015.
17. Nakamura et al. 2014.
18. Nishida 1983a.
19. Boesch et al. 2010.
20. Ibid.

21. Hobaiter, Schel, et al. 2014.
22. Pontzer and Wrangham 2006.
23. Pusey 1983, 1990.
24. For information on life expectancy in the wild, see Muller and Wrangham 2014 (Kanayawara); Hill et al. 2001 (Gombe and Taï); Nishida et al. 2003 (Mahale); Sugiyama et al. 1993; and Williams et al. 2008.
25. Muller and Wrangham 2014.
26. Wood et al. 2016.
27. Gillespie et al. 2010; Terio et al. 2011, 2016.
28. Hanamura et al. 2008.
29. Boesch et al. 2008.
30. Muller and Wrangham 2014.

7. Why Chimpanzees Hunt

1. Goodall 1963, 1968.
2. Teleki 1973.
3. This clearing is the so-called feeding station, a small, artificial clearing that eventually featured a cement storage box and dispensary for bananas and a couple of prefabricated metal observation huts. The space was created not only to provision the shy chimpanzees with bananas but also to allow filmmaker Hugo van Lawick to set up his bulky camera equipment close to the animals and avoid having to carry it all higher into the hills. Nearly all the early film footage of Gombe chimpanzee social behavior was shot here. Today the huts and bananas are long gone, but the chimpanzees still pass through the clearing regularly and socialize there, as a result of the behavioral traditions established so many decades ago.
4. Teleki 1973.
5. Busse 1977, 1978.
6. Teleki 1973.
7. Nishida, Uehara, and Nyundo 1979.
8. Boesch 1994a, 1994b, 1994c.
9. It is possible that adult and infant red colobus differ in taste, nutrient content, or some other factor, but this is difficult to test. This monkey species is widely eaten as bushmeat in West Africa, where it is considered to be one of the more flavorful species of monkey. Ethical concerns prevent us from supporting the bushmeat trade by obtaining a carcass to carry out nutritional studies. It's of course possible that whatever juvenile colobus monkeys lack in overall caloric content and nutrients, they make up for by being tender to eat.
10. Stanford 1996, 1998.
11. Wrangham and van Zinnicq Bergmann-Riss 1990.

12. Stanford 1996, 1998.
13. Stanford, Wallis, Matama, et al. 1994; Stanford, Wallis, Mpongo, et al. 1994. Chimpanzees have also no doubt influenced the evolution of antipredator defensive behavior in the red colobus monkeys, as well as impacted their demography and population biology. See Stanford 1995, 1999.
14. Pruetz and Bertolani 2007; Pruetz et al. 2015. The media took Pruetz's reports of tool use during hunting by Fongoli chimpanzees and portrayed it as evidence of spear making and spear use. Pruetz had not reported any such thing. The sticks are tapered by the chimps using their teeth, then used to ram galagos in order to injure or immobilize them before pulling them from their tree cavities by hand.
15. Speth and Davis 1976.
16. Stanford et al. 1994a; Stanford 1998; Gilby 2006.
17. Stanford 1998.
18. Watts and Mitani 2002.
19. Stanford 1998; Gilby et al. 2015.
20. Watts and Mitani 2002.
21. Stanford 1998.
22. Watts and Mitani 2002.
23. Fahy et al. 2013.
24. Watts 2008.
25. Archival footage from 1987, Jane Goodall Research Center, University of Southern California.
26. Newton-Fisher 1999.
27. Tennie, O'Malley, and Gilby 2014.
28. Uehara et al. 1992.
29. Muller and Lipson 2003; Muller and Wrangham 2004.
30. Watts and Mitani 2002.
31. Gilby and Wrangham 2007.
32. Wrangham et al. 1993, 1998.
33. Boesch 1994a.
34. Tennie, O'Malley, and Gilby 2014.
35. Teleki 1973.
36. Gilby et al. 2010.
37. Watts and Mitani 2002.
38. Gilby 2006; Gilby et al. 2010.
39. Watts and Mitani 2002; see also Stanford et al. 1994a, 1994b, 1995, and 1998.
40. Stanford et al. 1994b; Stanford 1998.
41. Gilby et al. 2010.
42. Boesch and Boesch 1989.
43. Gomes and Boesch 2009.

44. O'Malley et al. 2016.
45. Washburn and Lancaster 1968.

8. Got Culture?

1. McGrew 1992. See also McGrew 2004.
2. Wrangham et al. 2016.
3. Whiten et al. 1999. See also Whiten et al. 2001.
4. Nishida, Matsusaka, and McGrew 2009.
5. Hobaiter, Poisot, et al. 2014.
6. Gruber, Clay, and Zuberbühler 2010. See also Gruber and Clay 2016 for comparisons with bonobos.
7. Boesch and Boesch 1982.
8. Sirianni, Mundry, and Boesch 2015.
9. Arroyo et al. 2016.
10. Boesch and Boesch 1982
11. Matsuzawa 2006.
12. Inoue-Nakamura 1997.
13. Musgrave et al. 2016.
14. Schrauf et al. 2012.
15. Toth et al. 1993; Toth and Schick 2009.
16. Tomasello 1996.
17. Mercader, Panger, and Boesch 2002; Mercader et al. 2007.
18. Sanz, Schöning, and Morgan 2010. See also Sanz, Morgan, and Gulick 2004.
19. Hernandez-Aguilar 2009. See also Haslam et al. 2009.
20. Schöning et al. 2007. See also Schöning et al. 2008.
21. Stanford et al. 2000. Chimps in Bwindi were such avid honey foragers that we could identify the location of beehives high in trees by the presence of piles of broken sticks on the ground at the base of the trunk. Each stick was redolent of honey.
22. McLennan 2011, 2014.
23. Hicks, Fouts, and Fouts 2005.
24. McLennan 2015.
25. Boesch, Head, and Robbins 2009.
26. Gruber et al. 2009. See also Gruber et al. 2011.
27. Bogart and Pruetz 2011. See also Bogart and Pruetz 2008.
28. Lonsdorf 2005.
29. Lonsdorf, Anderson, 2014; Lonsdorf, Markham, et al. 2014.
30. O'Malley and Power 2012.
31. McGrew 2014.
32. Schöning et al. 2008.
33. Koops, Schöning, Isaji, et al. 2015; Koops, Schöning, McGrew, et al. 2015.

34. Schöning et al. 2008.
35. O'Malley et al. 2012.
36. Huffman and Kalunde 1993.
37. Pruetz and Bertolani 2007.
38. Pruetz et al. 2015.
39. Sanz, Morgan, and Hopkins 2016.
40. Lonsdorf and Hopkins 2005.
41. Humle and Matsuzawa 2009.
42. Kühl et al. 2016.
43. Ibid.
44. Lycett, Collard, and McGrew 2010. See also Lycett, Collard, and McGrew 2011.
45. Langergraber and Vigilant 2011.
46. Gruber et al. 2012.
47. Ibid.
48. Kamilar and Marshack 2012.

9. Blood Is Thicker

1. Tyson 1699 (1972).
2. Cited in Humle et al. 2016.
3. McBrearty and Jablonski 2005.
4. Sarich and Wilson 1967.
5. Sibley and Ahlquist 1984.
6. Ruvolo 1997.
7. Gagneux et al. 1999.
8. Goodman et al. 1998.
9. Chen and Li 2001.
10. Wilson and King 1975.
11. Varki and Atheide 2005.
12. Dorus et al. 2004.
13. Bustamante et al. 2005.
14. Ennard et al. 2002.
15. Rogers, Iltis, and Wooding 2004.
16. Finch and Stanford 2004. Many of us worry about consuming unhealthy amounts of saturated fat or cholesterol-rich food. Our paper suggested that we miss the big picture—that our bodies are far better able to process such a meaty diet than our closest kin. The implication of our paper that a mutation or mutations must have conferred the ability of our early hominin ancestors to include more meat in their diet without the attendant health risks has hopefully led researchers to search for such genes.
17. Keele et al. 2009.
18. Ibid.

19. Moorjani et al. 2016; Langergraber et al. 2012.
20. Wilson and King 1975.
21. Prüfer et al. 2015.
22. Prado-Martinez et al. 2013.
23. Sarmiento, Butynski, and Kalina 1996; Garner and Ryder 1996.
24. Gonder et al. 1997; Gonder et al. 2006. See also Chapter 1, note 3.
25. Fischer et al. 2011. See also Fischer et al. 2006.
26. Fischer et al. 2011, Fünfstück et al. 2015. Pigeonholing species and subspecies, as well as human racial variation, has a long history. Recent genetic studies show that species often have fuzzy borders, and that the gene pools composing them should be treated as dynamic rather than static entities.
27. Takemoto, Kawamoto, and Furuichi 2015.
28. Goldberg and Wrangham 1997.
29. Vigilant et al. 2001.
30. Inoue et al. 2008.
31. Langergraber, Rowney, Schubert, et al. 2014.
32. Patterson et al. 2006.
33. McCarthy et al. 2015.

10. Ape into Human

1. Darwin 1871.
2. Washburn 1968.
3. Johanson and White 1979.
4. Lovejoy 1978.
5. Stern and Susman 1983; Susman, Stern, and Jungers 1984.
6. Sayers and Lovejoy 2008; Lovejoy 2009; White et al. 2009, 2015.
7. See also Stanford 2012a.
8. Berger et al. 2015.
9. Dirks et al. 2015.
10. Val 2016.
11. Whiten et al. 1999, 2001.
12. Whiten, Schick, and Toth 2009. See also Toth et al. 1993; Toth and Schick 2009.
13. Haslam 2014. See also Haslam et al. 2009.
14. McGrew 2014.
15. D'Errico and Blackwell 2009.
16. Sanz, Morgan, and Gulick 2004; Sanz, Schöning, and Morgan 2010.
17. Pruetz 2007; Pruetz and Bertolani 2009.
18. Pruetz et al. 2015.
19. Kühl et al. 2016.
20. Stanford 2012b.

Bibliography

Alberts, S. C., J. C. Buchan, and J. Altmann. 2006. "Sexual selection in wild baboons: From mating opportunities to paternity success." *Animal Behaviour,* 72: 1177–1196.

Alberts, S. C., H. Watts, and J. Altmann. 2003. "Queuing and queue jumping: Long-term patterns of reproductive skew in male savannah baboons, *Papio cynocephalus.*" *Animal Behaviour,* 65: 821–840.

Anderson, M. J., S. J. Chapman, E. N. Videan, E. Evans, J. Fritz, T. S. Stoinski, A. F. Dixson, and P. Gagneux. 2007. "Functional evidence for differences in sperm competition in humans and chimpanzees." *American Journal of Physical Anthropology,* 134: 274–280.

Arnold, K., and A. Whiten. 2001. "Post-conflict behaviour of wild chimpanzees (*Pan troglodytes schweinfurthii*) in Budongo Forest, Uganda." *Behaviour,* 138: 649–690.

———. 2003. "Grooming interactions among the chimpanzees of the Budongo forest, Uganda: Tests of five explanatory models." *Behaviour,* 140: 519–552.

Arroyo, A., S. Hirata, T. Matsuzawa, and I. de la Torre. 2016. "Nut cracking tools used by captive chimpanzees (*Pan troglodytes*) and their comparison with Early Stone Age percussive artefacts from Olduvai Gorge." *PLoS ONE,* 11 (11): e0166788.

Atsalis, S., and E. Videan. 2009. "Reproductive aging in captive and wild common chimpanzees: Factors influencing the rate of follicular depletion." *American Journal of Primatology,* 71: 271–282.

Aureli, F., C. M. Schaffner, J. Verpooten, K. Slater, and G. Ramos-Fernandez. 2006. "Raiding parties of male spider monkeys: Insights into human warfare." *American Journal of Physical Anthropology,* 131: 486–497.

Babiszewska, M., S. A. Schel, C. Wilke, and K. E. Slocombe. 2015. "Social, contextual, and individual factors affecting the occurrence and acoustic structure of drumming bouts in wild chimpanzees (*Pan troglodytes*)." *American Journal of Physical Anthropology,* 156: 125–134.

Ban, S. D., C. Boesch, A. N'guessan, E. K. N'Goran, A. Tako, and K. R. L. Janmaat. 2016. "Taï chimpanzees change their travel direction for rare feeding trees providing fatty fruits." *Animal Behaviour,* 118: 135–147.

Basabose, A. K. 2004. "Fruit availability and chimpanzee party size at Kahuzi montane forest, Democratic Republic of Congo." *Primates,* 45: 211–219.

———. 2005. "Ranging patterns of chimpanzees in a montane forest of Kahuzi, Democratic Republic of Congo." *International Journal of Primatology,* 26: 33–53.

Bates, L. A., and R. W. Byrne. 2009. "Sex differences in the movement patterns of free-ranging chimpanzees (*Pan troglodytes schweinfurthii*): Foraging and border checking." *Behavioral Ecology and Sociobiology,* 64: 247–255.

Bearzi, M., and C. B. Stanford. 2007. "Dolphins and African apes: Comparisons of sympatric socio-ecology." *Contributions to Zoology,* 76: 235–254.

Benito-Calvo, A., S. Carvalho, A. Arroyo, T. Matsuzawa, and I. de la Torre. 2015. "First GIS analysis of modern stone tools used by wild chimpanzees (*Pan troglodytes verus*) in Bossou, Guinea, West Africa." *PLoS ONE,* 10 (3): e0121613.

Berger, L. R., J. Hawks, D. J. de Ruiter, S. E. Churchill, P. Schmid, L. K. Delezne, T. L. Kivell, et al. 2015. "*Homo naledi,* a new species of the genus *Homo* from the Dinaledi Chamber, South Africa." *eLife,* 4: e09560.

Bertolani, P., and J. D. Pruetz. 2011. "Seed reingestion in savannah chimpanzees (*Pan troglodytes verus*) at Fongoli, Senegal." *International Journal of Primatology,* 32: 1123–1132.

Boehm, C. 1994. "Pacifying interventions at Arnhem Zoo and Gombe." In *Chimpanzee Cultures,* edited by R. W. Wrangham, W. C. McGrew, F. B. M. de Waal, and P. G. Heltne, 211–226. Cambridge, Mass.: Harvard University Press.

———. 1999. *Hierarchy in the Forest: The Evolution of Egalitarian Behavior.* Cambridge, Mass.: Harvard University Press.

Boesch, C. 1994a. "Chimpanzees—red colobus monkeys: A predator-prey system." *Animal Behaviour,* 47: 1135–1148.

———. 1994b. "Cooperative hunting in wild chimpanzees." *Animal Behaviour,* 48: 653–667.

———. 1994c. "Hunting strategies of Gombe and Taï chimpanzees." In *Chimpanzee Cultures,* edited by R. W. Wrangham, W. C. McGrew, F. B. M. de Waal, and P. G. Heltne, 77–92. Cambridge, Mass.: Harvard University Press.

———. 1997. "Evidence for dominant wild female chimpanzees investing more in sons." *Animal Behaviour,* 54: 811–815.

Boesch, C., and H. Boesch 1982. "Optimization of nut-cracking with natural hammers by wild chimpanzees." *Behaviour,* 3: 265–286.

———. 1989. "Hunting behavior of wild chimpanzees in the Taï National Park." *American Journal of Physical Anthropology,* 78: 547–573.

Boesch, C., and H. Boesch-Achermann. 2000. *The Chimpanzees of the Taï Forest.* Oxford: Oxford University Press.

Boesch, C., C. Bolé, N. Eckhardt, and H. Boesch. 2010. "Altruism in forest chimpanzees: The case of adoption." *PLoS ONE,* 5 (1): e8901.

Boesch, C., C. Crockford, I. Herbinger, R. Wittig, Y. Moebius, and E. Normand. 2008. "Intergroup conflicts among chimpanzees in Taï National Park: Lethal violence and the female perspective." *American Journal of Primatology,* 70: 519–532.

Boesch, C., J. Head, and M. M. Robbins. 2009. "Complex tool sets for honey extraction among chimpanzees in Loango National Park, Gabon." *Journal of Human Evolution,* 56: 560–569.

Boesch, C., G. Kohou, H. Néné, and L. Vigilant. 2006. "Male competition and paternity in wild chimpanzees of the Taï forest." *American Journal of Physical Anthropology,* 130: 103–115.

Bogart, S. L., and J. D. Pruetz. 2008. "Ecological context of savanna chimpanzee (*Pan troglodytes verus*) termite fishing at Fongoli, Senegal." *American Journal of Primatology,* 70: 605–612.

———. 2011. "Insectivory of savanna chimpanzees (*Pan troglodytes verus*) at Fongoli, Senegal." *American Journal of Physical Anthropology,* 145: 11–20.

Botero, M., S. E. MacDonald, and R. S. Miller. 2013. "Anxiety-related behavior of orphan chimpanzees (*Pan troglodytes schweinfurthii*) at Gombe National Park, Tanzania." *Primates,* 54: 21–26.

Bradley, B. J., M. M. Robbins, E. A. Williamson, H. D. Steklis, N. Gerald Steklis, N. Eckhardt, C. Boesch, and L. Vigilant. 2005. "Mountain gorilla tug-of-war: Silverbacks have limited control over reproduction in multimale groups." *Proceedings of the National Academy of Sciences,* 102: 9418–9423.

Brindle, M., and C. Opie. 2016. "Postcopulatory sexual selection influences baculum evolution in primates and carnivores." *Proceedings of the Royal Society B (Biological Sciences),* 283: 201617361.

Busse, C. D. 1977. "Chimpanzee predation as a possible factor in the evolution of red colobus monkey social organization." *Evolution,* 31: 907–911.

———. 1978. "Do chimpanzees hunt cooperatively?" *American Naturalist,* 112: 767–770.

Bustamante, C., A. Fledel-Alon, S. Willamson, R. Nielsen, M. Todd Hubisz, S. Glanowski, D. M. Tanenbaum, et al. 2005. "Natural selection on protein-coding genes in the human genome." *Nature,* 437: 1153–1157.

Bygott, D. 1979. "Agonistic behavior and dominance among wild chimpanzees." In *The Great Apes,* edited by D. Hamburg and E. McCown, 405–427. Menlo Park, Calif.: B. Cummings.

Carvalho, J. S., L. Vicente, and T. A. Marques. 2015. "Chimpanzee (*Pan troglodytes verus*) diet composition and food availability in a human-modified landscape at Lagoas de Cufada Natural Park, Guinea-Bissau." *International Journal of Primatology*, 36: 802–822.

Chancellor, R. L., A. S. Rundus, and S. Nyandwi. 2012. "The influence of seasonal variation on chimpanzee (*Pan troglodytes schweinfurthii*) fallback food consumption, nest group size, and habitat use in Gishwati, a montane rain forest fragment in Rwanda." *International Journal of Primatology*, 33: 115–133.

Chapman, C. A., F. J. White, and R. W. Wrangham. 1994. "Party size in chimpanzees and bonobos." In *Chimpanzee Cultures*, edited by R. W. Wrangham, W. C. McGrew, F. B. M. de Waal, and P. Heltne, 41–57. Cambridge, Mass.: Harvard University Press.

Check, E. 2004. "Geneticists study chimp-human divergence." *Nature*, 428: 242.

Chen, F., and W. Li. 2001. "Genomic divergence between humans and other hominoids and the effective population size of the common ancestor of humans and chimpanzees." *American Journal of Human Genetics*, 68: 444–456.

Clark Arcadi, A., D. Robert, and C. Boesch. 1998. "Buttress drumming by wild chimpanzees: Temporal patterning, phase integration into loud calls, and preliminary evidence for individual distinctiveness." *Primates*, 39: 505–518.

Clark Arcadi, A., and R. W. Wrangham. 1999. "Infanticide in chimpanzees: Review of cases and a new within-group observation from the Kanyawara study group in Kibale National Park." *Primates*, 40: 337–351.

Clutton-Brock, T. H., and P. H. Harvey. 1977. "Primate ecology and social organisation." *Journal Zoology (London)*, 183: 1–39.

Creel, S. 2001. "Social dominance and stress hormones." *Trends in Ecology and Evolution*, 16: 491–497.

Crockford, C., R. M. Wittig, R. Mundry, and K. Zuberbühler. 2012. "Wild chimpanzees inform ignorant group members of danger." *Current Biology*, 22: 142–146.

Crook, J. H., and S. J. Gartlan. 1966. "On the evolution of primate societies." *Nature*, 210: 1200–1203.

Darwin, C. 1871. *The Descent of Man and Selection in Relation to Sex*. London: J. Murray.

D'Errico, F., and L. Blackwell. 2009. "Assessing the function of early hominin bone tools." *Journal of Archaeological Science*, 36: 1764–1773.

De Ruiter, J. R., W. Scheffrahn, G. Trommelen, A. Uitterlinden, R. Martin, and J. Van Hoof. 1992. "Male social rank and reproductive success in wild long-tailed macaques." In *Paternity in Primates: Genetic Tests and*

Theories, edited by R. Martin, A. Dixson, and E. Wickings, 175–191. Basel, Switzerland: Karger.

De Ruiter, J. R., and J. A. R. A. M. van Hooff. 1993. "Male dominance rank and reproductive success in primate groups." *Primates,* 34: 513–523.

Deschner, T., M. Heistermann, K. Hodges, and C. Boesch. 2004. "Female swelling size, timing of ovulation, and male behavior in wild West African chimpanzees." *Hormones and Behavior,* 46, 204–215.

DeVore, I. 1965. "Male dominance and mating behavior in baboons." In *Sex and Behavior,* edited by F. Beach, 266–290. New York: Wiley.

De Waal, F. B. M. 1982. *Chimpanzee Politics: Power and Sex among Apes.* Baltimore: Johns Hopkins University Press.

———. 2001. *The Ape and the Sushi Master.* New York: Basic Books.

De Waal, F. B. M., and F. Aureli. 1997. "Conflict resolution and distress alleviation in monkeys and apes." *Annals of the New York Academy of Sciences,* 807: 317–328.

De Waal, F. B. M., and L. Luttrell. 1988. "Mechanisms of social reciprocity in three primate species: Symmetrical relationship characteristics or cognition?" *Ethology and Sociobiology,* 9: 101–118.

De Waal, F. B. M., and M. van Roosmalen. 1979. "Reconciliation and consolation among chimpanzees." *Behavioral Ecology and Sociobiology,* 5: 55–66.

Dirks, P. H. G. M., L. R. Berger, E. M. Roberts, J. D. Kramers, J. Hawks, P. S. Randolph-Quinney, M. Elliott, et al. 2015. "Geological and taphonomic context for the new hominin species *Homo naledi* from the Dinaledi Chamber, South Africa." *eLife,* 4: e09561.

Dixson, A. F. 2012. *Primate Sexuality: Comparative Studies of the Prosimians, Monkeys, Apes and Humans.* 2nd ed. Oxford: Oxford University Press.

Dixson, A. F., and N. I. Mundy. 1994. "Sexual behavior, sexual swelling, and penile evolution in chimpanzees (*Pan troglodytes*)." *Archives of Sexual Behavior,* 23: 267–280.

Dorus, S., E. J. Vallender, P. D. Evans, J. R. Anderson, S. L. Gilbert, M. Mahowald, G. J. Wyckoff, C. M. Malcolm, and B. T. Lahn. 2004. "Accelerated evolution of nervous system genes in the origin of *Homo sapiens.*" *Cell,* 119: 1027–1040.

Duffy, K. G., R. W. Wrangham, and J. B. Silk. 2007. "Male chimpanzees exchange political support for mating opportunities." *Current Biology,* 17: R586–R587.

Dutton, P., and H. Chapman. 2015. "New tools suggest local variation in tool use by a montane community of the rare Nigeria-Cameroon chimpanzee, *Pan troglodytes ellioti,* in Nigeria." *Primates,* 56: 89–100.

Ely, J. J., W. I. Frels, S. Howell, M. K. Izard, M. E. Keeling, and D. R. Lee. 2006. "Twinning and heteropaternity in chimpanzees (*Pan troglodytes*)." *American Journal of Physical Anthropology,* 130: 96–102.

Emery, M. A., and P. L. Whitten. 2003. "Size of sexual swellings reflects ovarian function in chimpanzees (*Pan troglodytes*)." *Behavioral Ecology and Sociobiology*, 54: 340–351.

Emery Thompson, M. 2005. "Reproductive endocrinology of wild female chimpanzees (*Pan troglodytes schweinfurthii*): Methodological considerations and the role of hormones in sex and conception." *American Journal of Primatology*, 67: 137–158.

———. 2013. "Reproductive ecology of female chimpanzees." *American Journal of Primatology*, 75: 222–237.

———. 2014. "Sexual conflict: Nice guys finish last." *Current Biology*, 24: R1125–R1126.

Emery Thompson, M., J. H. Jones, A. E. Pusey, S. Brewer-Marsden, J. Goodall, D. Marsden, T. Matsuzawa, et al. 2007. "Aging and fertility patterns in wild chimpanzees provide insights into the evolution of menopause." *Current Biology*, 17: 2150–2156.

Emery Thompson, M., M. N. Muller, K. Sabbi, Z. P. Machanda, E. Otali, and R. W. Wrangham. 2016. "Faster reproductive rates trade off against offspring growth in wild chimpanzees." *Proceedings of the National Academy of Sciences*, 113: 7780–7785.

Emery Thompson, M., M. N. Muller, and R. W. Wrangham. 2014. "Male chimpanzees compromise the foraging success of their mates in Kibale National Park, Uganda." *Behavioral Ecology and Sociobiology*, 68: 1973–1983.

Emery Thompson, M., N. Newton-Fisher, and V. Reynolds. 2006. "Probable community transfer of parous adult female chimpanzees in the Budongo Forest, Uganda." *International Journal of Primatology*, 1601–1617.

Emery Thompson, M., and R. W. Wrangham. 2008. "Diet and reproductive function in wild female chimpanzees (*Pan troglodytes schweinfurthii*) at Kibale National Park, Uganda." *American Journal of Physical Anthropology*, 135: 171–181.

Ennard, W., M. Przeworski, S. E. Fisher, C. S. Lai, V. Wiebe, T. Kitano, A. P. Monaco, and S. Pääbo. 2002. "Molecular evolution of FOXP2, a gene involved in speech and language." *Nature*, 418: 869–872.

Fahy, G. E., M. Richards, J. Riedel, J. J. Hublin, and C. Boesch. 2013. "Stable isotope evidence of meat eating and hunting specialization in adult male chimpanzees." *Proceedings of the National Academy of Sciences*, 110: 5829–5833.

Fallon, B. L., C. Newmann, R. W. Byrne, and K. Zuberbühler. 2016. "Female chimpanzees adjust copulation calls according to reproductive status and level of female competition." *Animal Behaviour*, 113: 87–92.

Fanshawe, J. H. and C. D. Fitzgibbon (1993). "Factors influencing the hunting success of an African wild dog pack." *Animal Behaviour*, 45: 479–490.

Fawcett, K. A. 2000. "Female relationships and food availability in a forest community of chimpanzees." PhD diss., University of Edinburgh.

Fawcett, K. A., and G. Muhumuza. 2000. "Death of a wild chimpanzee community member: Possible outcome of intense sexual competition?" *American Journal of Primatology*, 51: 243–247.

Fedurek, P., E. Donnellan, and K. E. Slocombe. 2014. "Social and ecological correlates of long-distance pant hoot calls in male chimpanzees." *Behavioral Ecology and Sociobiology*, 68: 1345–1355.

Fedurek, P., Z. P. Machanda, A. M. Schel, and K. E. Slocombe. 2013. "Pant hoot chorusing and social bonds in male chimpanzees." *Animal Behaviour*, 86: 189–196.

Fedurek, P., K. E. Slocombe, D. E. Enigk, M. Emery Thompson, R. W. Wrangham, and M. N. Muller. 2016. "The relationship between testosterone and long-distance calling in wild male chimpanzees." *Behavioral and Ecology and Sociobiology*, 70: 659–679.

Fedurek, P., K. E. Slocombe, and K. Zuberbühler. 2015. "Chimpanzees communicate to two different audiences during aggressive interactions." *Animal Behaviour*, 110: 21–28.

Fedurek, P., K. Zuberbühler, and C. D. Dahl. 2016. "Sequential information in a great ape utterance." *Scientific Reports*, 6: 38226.

Feldblum, J. T., C. Krupenye, E. E. Wroblewski, R. S. Rudicell, B. H. Hahn, A. E. Pusey, and I. C. Gilby. 2015. "The adaptive value of male relationships in the chimpanzees of Gombe National Park, Tanzania." *American Journal of Physical Anthropology*, 156 (S60): 132 (abstract).

Feldblum, J. T., E. E. Wroblewski, R. R. Rudicell, B. H. Hahn, T. Paiva, M. Cetinkaya-Rundel, A. E. Pusey, and I. C. Gilby. 2014. "Sexually coercive male chimpanzees sire more offspring." *Current Biology*, 24: 2855–2860.

Finch, C. E., and C. B. Stanford. 2004. "Meat-adaptive genes and the evolution of slower aging in humans." *Quarterly Review of Biology*, 79: 1–50.

Fischer, A., J. Pollack, O. Thalmann, B. Nickel, and S. Pääbo. 2006. "Demographic history and genetic differentiation in apes." *Current Biology*, 16: 1133–1138.

Fischer, A., K. Prüfer, J. M. Good, M. Halbwax, V. Wiebe, C. André, R. Atencia, L. Mugisha, S. E. Ptak, and S. Pääbo. 2011. "Bonobos fall within the genomic variation of chimpanzees." *PLoS ONE*, 6 (6): e21605.

Foerster, S., M. Franz, C. M. Murray, I. C. Gilby, J. T. Feldblum, K. K. Walker, and A. E. Pusey. 2016. "Chimpanzee females queue but males compete for social status." *Scientific Reports*, 6: 35404.

Foerster, S., K. McLellan, K. Schroepfer-Walker, C. M. Murray, C. Krupenye, I. C. Gilby, and A. E. Pusey. 2015. "Social bonds in the dispersing sex: Partner preferences among adult female chimpanzees." *Animal Behaviour*, 105: 139–152.

Foster, M. W., I. C. Gilby, C. M. Murray, A. Johnson, E. E. Wroblewski, and A. E. Pusey. 2009. "Alpha male chimpanzee grooming patterns: Implications for dominance 'style.'" *American Journal of Primatology,* 71: 136–144.

Fowler, A., Y. Koutsioni, and V. Sommer. 2007. "Leaf-swallowing in Nigerian chimpanzees: Evidence for assumed self-medication." *Primates,* 48: 73–76.

Fowler, A., and V. Sommer. 2007. "Subsistence technology in Nigerian chimpanzees." *International Journal of Primatology,* 28: 997–1023.

Fröhlich, M., R. M. Wittig, and S. Pika. 2016. "Should I stay or should I go? Initiation of joint travel in mother-infant dyads of two chimpanzee communities in the wild." *Animal Cognition,* 19 (3): 483–500.

Fry, D. P., and P. Söderberg. 2013. "Lethal aggression in mobile forager bands and implications for the origins of war." *Science,* 341: 270–273.

Fujisawa, M., K. J. Hockings, A. G. Soumah, and T. Matsuzawa. 2016. "Placentophagy in wild chimpanzees (*Pan troglodytes verus*) at Bossou, Guinea." *Primates,* 57: 175–180.

Fünfstück, T., M. Arandjelovic, D. B. Morgan, C. Sanz, P. Reed, S. H. Olson, K. Cameron, A. Ondzie, M. Peeters, and L. Vigilant. 2015. "The sampling scheme matters: *Pan troglodytes troglodytes* and *P. t. schweinfurthii* are characterized by clinal genetic variation rather than a strong subspecies break." *American Journal of Physical Anthropology,* 156: 181–191.

Funston, P. J., M. G. L. Mills, and H. C. Biggs. 2001. "Factors affecting the hunting success of male and female lions in the Kruger National Park." *Journal of Zoology,* 253: 419–431.

Furuichi, T. 2009. "Factors underlying party size differences between chimpanzees and bonobos: A review and hypotheses for future study." *Primates,* 50: 197–209.

Gagneux, P., C. Wills, U. Gerloff, D. Tautz, P. A. Morin, C. Boesch, B. Fruth, G. Hohmann, O. A. Ryder, and D. S. Woodruff. 1999. "Mitochondrial sequences show diverse evolutionary histories of African hominoids." *Proceedings of the National Academy of Sciences,* 96: 5077–5082.

Garner, K. J., and O. A. Ryder. 1996. "Mitochondrial DNA diversity in gorillas." *Molecular Phylogenetics and Evolution,* 6: 39–48.

Georgiev, A. V., A. F. Russell, M. E. Thompson, E. Otali, M. N. Muller, and R. W. Wrangham. 2014. "The foraging costs of mating effort in male chimpanzees (*Pan troglodytes schweinfurthii*)." *International Journal of Primatology,* 35: 725–745.

Ghiglieri, M. P. 1987. "Sociobiology of the great apes and the hominid ancestor." *Journal of Human Evolution,* 16: 319–357.

Gibbons, A. 1998. "Comparative genetics: Which of our genes make us human?" *Science,* 281: 1432–1434.

Gilby, I. C. 2006. "Meat sharing among the Gombe chimpanzees: Harassment and reciprocal exchange." *Animal Behaviour,* 71: 953–963.

Gilby, I. C., L. J. N. Brent, E. E. Wroblewski, R. S. Rudicell, B. H. Hahn, J. Goodall, and A. E. Pusey. 2013. "Fitness benefits of coalitionary aggression in male chimpanzees." *Behavioral Ecology and Sociobiology*, 67: 373–381.

Gilby, I. C., and R. C. Connor. 2010. "The role of intelligence in group hunting: Are chimpanzees different from other social predators?" In *The Mind of the Chimpanzee: Ecological and Experimental Perspectives*, edited by E. V. Lonsdorf, S. R. Ross, and T. Matsuzawa, 220–233. Chicago: University of Chicago Press.

Gilby, I. C., M. Emery Thompson, J. Ruane, and R. W. Wrangham. 2010. "No evidence of short-term exchange of meat for sex among chimpanzees." *Journal of Human Evolution*, 59: 44–53.

Gilby, I. C., Z. P. Machanda, D. C. Mjungu, J. Rosen, M. N. Muller, A. E. Pusey, and R. W. Wrangham. 2015. "Impact hunters catalyse cooperative hunting in two wild chimpanzee communities." *Philosophical Transactions of the Royal Society Series B*, 370: 20150005.

Gilby, I. C., M. L. Wilson, and A. E. Pusey. 2013. "Ecology rather than psychology explains co-occurrence of predation and border patrols in male chimpanzees." *Animal Behaviour*, 86: 61–74.

Gilby, I. C., and R. W. Wrangham. 2007. "Risk-prone hunting by chimpanzees (*Pan troglodytes schweinfurthii*) increases during periods of high diet quality." *Behavioral Ecology and Sociobiology*, 61: 1771–1779.

Gillespie, T. R., E. V. Lonsdorf, E. P. Canfield, D. J. Meyer, Y. Nadler, J. Raphael, A. E. Pusey, et al. 2010. "Demographic and ecological effects on patterns of parasitism in eastern chimpanzees (*Pan troglodytes schweinfurthii*) in Gombe National Park, Tanzania." *American Journal of Physical Anthropology*, 143: 534–544.

Glazko, G., V. Veeramachaneni, M. Nei, and W. Makalowski. 2005. "Eighty percent of proteins are different between humans and chimpanzees." *Gene*, 346: 215–219.

Glowacki, L., A. Isakov, R. W. Wrangham, R. McDermott, J. H. Fowler, and N. A. Christakis. 2016. "Formation of raiding parties for intergroup violence mediated by social network structure." *Proceedings of the National Academy of Sciences*, 113: 12114–12119.

Goldberg, A., and R. W. Wrangham. 1997. "Genetic correlates of social behaviour in chimpanzees: Evidence from mitochondrial DNA." *Animal Behaviour*, 54: 559–570.

Gomes, C. M., and C. Boesch. 2009. "Wild chimpanzees exchange meat for sex on a long-term basis." *PLoS ONE*, 4 (4): e5116.

Gomes, C. M., R. Mundry, and C. Boesch. 2009. "Long-term reciprocation of grooming in wild West African chimpanzees." *Proceedings of the Royal Society B (Biological Sciences)*, 276: 699–706.

Gonder, M. K., T. R. Disotell, and J. F. Oates. 2006. "New genetic evidence on the evolution of chimpanzee populations and implications for taxonomy." *International Journal of Primatology,* 27: 1103–1127.

Gonder, M. K., J. F. Oates, T. R. Disotell, M. R. Forstner, J. C. Morales, and D. J. Melnick. 1997. "A new West African chimpanzee subspecies?" *Nature,* 388: 337.

Goodall, J. 1963. "Feeding behaviour of wild chimpanzees: A preliminary report." *Symposia of the Zoological Society of London,* 10: 39–48.

———. 1968. "Behaviour of free-living chimpanzees of the Gombe Stream area." *Animal Behaviour Monographs,* 1:163–311.

———. 1971. *In the Shadow of Man.* With H. van Lawick. Boston: Houghton Mifflin.

———. 1986. *The Chimpanzees of Gombe: Patterns of Behavior.* Cambridge, Mass.: Harvard University Press.

Goodman, M., C. A. Porter, J. Czelusniak, S. L. Page, H. Schneider, J. Shoshani, G. Gunnell, and C. P. Groves. 1998. "Toward a phylogenetic classification of primates based on DNA evidence complemented by fossil evidence." *Molecular Phylogenetics and Evolution,* 9: 585–598.

Graham, C. E. 1979. "Reproductive function in aged female chimpanzees." *American Journal of Physical Anthropology,* 50: 291–300.

Gross-Camp, N. D., M. Masozera, and B. A. Kaplin. 2009. "Chimpanzee seed dispersal quantity in a tropical montane forest of Rwanda." *American Journal of Primatology,* 71: 901–911.

Gruber, T., and Z. Clay. 2016. "A comparison between bonobos and chimpanzees: A review and update." *Evolutionary Anthropology,* 25: 239–252.

Gruber, T., Z. Clay, and K. Zuberbühler. 2010. "A comparison of bonobo and chimpanzee tool use: Evidence for a female bias in the *Pan* lineage." *Animal Behaviour,* 80: 1023–1033.

Gruber, T., M. N. Muller, V. Reynolds, R. Wrangham, and K. Zuberbühler. 2011. "Community-specific evaluation of tool affordances in wild chimpanzees." *Scientific Reports,* 1: 128.

Gruber, T., M. N. Muller, P. Strimling, R. Wrangham, and K. Zuberbühler. 2009. "Wild chimpanzees rely on cultural knowledge to solve an experimental honey acquisition task." *Current Biology,* 19: 1806–1810.

Gruber, T., K. B. Potts, C. Krupenye, M. Byrne, C. Mackworth-Young, W. C. McGrew, V. Reynolds, and K. Zuberbühler. 2012. "The influence of ecology on chimpanzee (*Pan troglodytes*) cultural behavior: A case study of five Ugandan chimpanzee communities." *Journal of Comparative Psychology,* 126: 446–457.

Gruber, T., K. Zuberbühler, and C. Newmann. 2016. "Travel fosters tool use in wild chimpanzees." *eLife,* 5: e16371.

Grueter, C. C. 2015. "Home range overlap as a driver of intelligence in primates." *American Journal of Primatology,* 77: 418–424.

Haile-Selassie, Y., B. M. Latimer, M. Alene, A. L. Deino, L. Gibert, S. M. Melillo, B. Z. Saylor, G. R. Scott, and C. O. Lovejoy. 2010. "An early Australopithecus afarensis postcranium from Woranso-Mille, Ethiopia." *Proceedings of the National Academy of Sciences,* 107: 12121–12126.

Haley, M. P., C. J. Deutsch, and B. J. Le Boeuf. 1994. "Size, dominance and copulatory success in male northern elephant seals, *Mirounga angustirostris.*" *Animal Behaviour,* 48: 1249–1260.

Hanamura, S., M. Kiyono, M. Lukasik-Braum, T. Mlengeya, M. Fujimoto, M. Nakamura, and T. Nishida. 2008. "Chimpanzee deaths at Mahale caused by a flu-like disease." *Primates,* 49: 77–80.

Harcourt, A. H., P. H. Harvey, S. G. Larson, and R. V. Short. 1981. "Testis size, body weight and breeding system in primates." *Nature,* 293: 55–57.

Harcourt, A. H., and K. J. Stewart. 1987. "The influence of help in contests on dominance rank in primates: Hints from gorillas." *Animal Behaviour,* 35: 182–190.

Hartel, J. A. 2015. "Social dynamics of intragroup aggression and conflict resolution in wild chimpanzees (*Pan troglodytes*) at Kanyawara, Kibale National Park, Uganda." PhD diss., University of Southern California.

Hasegawa, T., M. Hiraiwa, T. Nishida, and H. Takasaki. 1983. "New evidence on scavenging behavior in wild chimpanzees." *Current Anthropology,* 24: 231–232.

Hashimoto, C., T. Furuichi, and T. Tashiro. 2001. "What factors affect the size of chimpanzee parties in the Kalinzu Forest, Uganda? Examination of fruit abundance and number of estrous females." *International Journal of Primatology,* 22: 947–959.

Hashimoto, C., M. Isaji, K. Koops, and T. Furuichi. 2015. "First records of tool-set use for ant-dipping by eastern chimpanzees (*Pan troglodytes schweinfurthii*) in the Kalinzu Forest, Uganda." *Primates,* 56: 301–305.

Hashimoto, C., S. Suzuki, Y. Takenoshita, J. Yamagiwa, A. K. Basabose, and T. Furuichi. 2003. "How fruit abundance affects the chimpanzee party size: A comparison between four study sites." *Primates,* 44: 77–81.

Haslam, M. 2014. "On the tool use behavior of the bonobo-chimpanzee last common ancestor, and the origins of hominine stone tool use." *American Journal of Primatology,* 76: 910–918.

Haslam, M., A. Hernandez-Aguilar, V. Ling, S. Carvalho, I. de la Torre, A. DeStefano, A. Du, et al. 2009. "Primate archaeology." *Nature,* 460: 339–344.

Hausfater, G. 1975. *Dominance and Reproduction in Baboons (*Papio cynocephalus*)*. Contributions to Primatology, vol. 7. Basel, Switzerland: Karger.

Hawkes, K. 2004. "The grandmother effect." *Nature,* 428: 128.

Hawkes, K., and K. R. Smith. 2010. "Do women stop early? Similarities in fertility decline in humans and chimpanzees." *Annals of the New York Academy of Sciences,* 1204: 43–53.

Hayakawa, T. 2015. "Taste of chimpanzee foods." In *Mahale Chimpanzees: 50 Years of Research,* edited by M. Nakamura, K. Hosaka, N. Itoh, and K. Zamma, 246–258. Cambridge: Cambridge University Press.

Heintz, M. R. 2013. "The immediate and long-term benefits of social play in wild chimpanzees (*Pan troglodytes*)." PhD diss., University of Chicago.

Hernandez-Aguilar, R. A. 2009. "Chimpanzee nest distribution and site reuse in a dry habitat: Implications for early hominin ranging." *Journal of Human Evolution,* 57: 350–364.

Hernandez-Aguilar, R. A., J. Moore, and T. R. Pickering. 2007. "Savanna chimpanzees use tools to harvest the underground storage organs of plants." *Proceedings of the National Academy of Sciences,* 104: 19210–19123.

Hernandez-Aguilar, R. A., J. Moore, and C. B. Stanford. 2013. "Chimpanzee nesting patterns in savanna habitat: Environmental influences and preferences." *American Journal of Primatology,* 75: 979–994.

Hicks, T. C., R. S. Fouts, and D. H. Fouts. 2005. "Chimpanzee (*Pan troglodytes troglodytes*) tool use in the Ngotto Forest, Central African Republic." *American Journal of Primatology,* 65: 221–237.

Hill, K., C. Boesch, J. Goodall, A. Pusey, J. Williams, and R. W. Wrangham. 2001. "Mortality rates among wild chimpanzees." *Journal of Human Evolution,* 40: 437–450.

Hinde, K., and L. A. Milligan. 2011. "Primate milk: Proximate mechanisms and ultimate perspectives." *Evolutionary Anthropology,* 20: 9–23.

Hobaiter, C., T. Poisot, K. Zuberbühler, W. Hoppit, and T. Gruber. 2014. "Social network analysis shows direct evidence for social transmission of tool use in wild chimpanzees." *PLoS Biology,* 12 (9): e1001960.

Hobaiter, C., A. M. Schel, K. Langergraber, and K. Zuberbühler. 2014. "'Adoption' by maternal siblings in wild chimpanzees." *PLoS ONE,* 9 (8): e103777.

Holekamp, K. E., L. Smale, R. Berg, and S. M. Cooper. 1997. "Hunting rates and hunting success in the spotted hyena (*Crocuta crocuta*)." *Journal of Zoology,* 242: 1–15.

Horner, V., D. Proctor, K. E. Bonnie, A. Whiten, and F. B. M. de Waal. 2010. "Prestige affects cultural learning in chimpanzees." *PLoS ONE,* 5 (5): e10625.

Hosaka, K., T. Nishida, M. Hamai, A. Matsumoto-Oda, and S. Uehara. 2001. "Predation of mammals by the chimpanzees of the Mahale Mountains, Tanzania." In *All Apes Great and Small,* vol. 1, *African Apes,* edited by B. M. F. Galdikas, N. E. Briggs, L. K. Sheeran, G. L. Shapiro, and J. Goodall, 107–130. Developments in Primatology: Progress and Prospects. New York: Kluwer.

Houle, A., C. Chapman, and W. L. Vickery. 2010. "Intratree vertical variation of fruit density and the nature of contest competition in frugivores." *Behavioral Ecology and Sociobiology,* 64: 429–441.

Hrdy, S. B. 1977. *The Langurs of Abu.* Cambridge, Mass.: Harvard University Press.

———. 1979. "Infanticide among animals: A review, classification, and examination of the implications for the reproductive strategy of females." *Ethology and Sociobiology,* 1: 13–40.

Huffman, M. A. "Chimpanzee self-medication: A historical perspective of the key findings." In *Mahale Chimpanzees: 50 Years of Research,* edited by M. Nakamura, K. Hosaka, N. Itoh, and K. Zamma, 340–353. Cambridge: Cambridge University Press.

Huffman, M. A., and M. S. Kalunde. 1993. "Tool-assisted predation on a squirrel by a female chimpanzee in the Mahale Mountains, Tanzania." *Primates,* 34: 93–98.

Hughes, J. F., H. Skaletsky, T. Pyntikova, T. A. Graves, S. K. van Daalen, P. J. Minx, R. S. Fulton, et al. 2015. "Chimpanzee and human Y chromosomes are remarkably divergent in structure and gene content." *Nature,* 463: 536–539.

Humle, T., F. Maisels, J. F. Oates, A. Plumptre, and E. A. Williamson. 2016. Pan troglodytes: *The IUCN Red List of Threatened Species 2016.* Cambridge, UK: International Union for Conservation of Nature and Natural Resources. http://dx.doi.org/10.2305/IUCN.UK.2016-2.RLTS.T15933A17964454.en.

Humle, T., and T. Matsuzawa. 2002. "Ant-dipping among the chimpanzees of Bossou, Guinea, and some comparisons with the other sites." *American Journal of Primatology,* 58: 133–148.

———. 2009. "Laterality in hand use across four tool-use behaviors among the wild chimpanzees of Bossou, Guinea, West Africa." *American Journal of Primatology,* 71: 40–48.

Inoue, E., M. Inoue-Murayama, L. Vigilant, O. Takenaka, and T. Nishida. 2008. "Relatedness in wild chimpanzees: Influence of paternity, male philopatry, and demographic factors." *American Journal of Physical Anthropology,* 137: 256–262.

Inoue-Nakamura, N. 1997. "Development of stone tool use by wild chimpanzees (*Pan troglodytes*)." *Journal of Comparative Psychology,* 111: 159–173.

Itoh, N., and T. Nishida. 2007. "Chimpanzee grouping patterns and food availability in Mahale Mountains National Park, Tanzania." *Primates,* 48: 87–96.

Janmaat, K. R. L., S. D. Ban, and C. Boesch. 2013. "Chimpanzees use long-term spatial memory to monitor large fruit trees and remember feeding experiences across seasons." *Animal Behaviour,* 86: 1183–1205.

Johanson, D. C., and T. D. White. 1979. "A systematic assessment of early African hominids." *Science,* 202: 321–330.

Jones, J. H., M. L. Wilson, C. Murray, and A. Pusey. 2010. "Phenotypic quality influences fertility in Gombe chimpanzees." *Journal of Animal Ecology,* 79: 1262–1269.

Kaburu, S. S. K., and N. E. Newton-Fisher. 2015. "Egalitarian despots: Hierarchy steepness, reciprocity and the grooming-trade model in wild chimpanzees, Pan troglodytes." *Animal Behaviour,* 99: 61–71.

Kachel, A. F., L. S. Premo, and J. J. Hublin 2011. "Grandmothering and natural selection." *Proceedings of the Royal Society B (Biological Sciences),* 278: 384–391.

Kahlenberg, S. M., M. Emery Thompson, M. N. Muller, and R. W. Wrangham. 2008. "Immigration costs for female chimpanzees and male protection as an immigrant counterstrategy to intrasexual aggression." *Animal Behaviour,* 76: 1497–1509.

Kahlenberg, S. M., and R. W. Wrangham. 2010. "Sex differences in chimpanzees' use of sticks as play objects resembles those of children." *Current Biology,* 20: R1067–R1068.

Kalan, A. K., and C. Boesch. 2015. "Audience effects in chimpanzee food calls and their potential for recruiting others." *Behavioral Ecology and Sociobiology,* 69: 1701–1712.

Kalan, A. K., R. Mundry, and C. Boesch. 2015. "Wild chimpanzees modify food call structure with respect to tree size for a particular fruit species." *Animal Behaviour,* 101: 1–9.

Kalcher-Sommersguter, E., S. Preuschoft, C. Franz-Schaider, C. K. Hemelrijk, K. Crailsheim, and J. J. M. Massen. 2015. "Early maternal loss affects social integration of chimpanzees throughout their lifetime." *Scientific Reports,* 5 (November 10): 16439.

Kamilar, J. M., and J. L. Marshack. 2012. "Does geography or ecology best explain 'cultural' variation among chimpanzee communities?" *Journal of Human Evolution,* 62: 256–260.

Keele, B. F., J. H. Jones, K. A. Terio, J. D. Estes, R. S. Rudicell, M. L. Wilson, Y. Li, et al. 2009. "Increased mortality and AIDS-like immunopathology in wild chimpanzees infected with SIVcpz." *Nature,* 460: 515–519.

Kehrer-Sawatzki, H., and D. N. Cooper. 2007. "Understanding the recent evolution of the human genome: Insights from human-chimpanzee genome comparisons." *Human Mutation,* 28: 99–130.

King, M. C., and A. C. Wilson. 1975. "Evolution at two levels in humans and chimpanzees." *Science,* 188: 107–116.

Kirk, J. 2011. *Kingdom under Glass: A Tale of Obsession, Adventure, and One Man's Quest to Preserve the World's Great Animals.* New York: Picador.

Koops, K., T. Furuichi, C. Hashimoto, and C. P. van Schaik. 2015. "Sex differences in object manipulation in wild immature chimpanzees (*Pan troglodytes schweinfurthii*) and bonobos (*Pan paniscus*): Preparation for tool use?" *PLoS ONE,* 10 (10): e0139909.

Koops, K., T. Humle, E. Sterck, and T. Matsuzawa. 2006. "Ground-nesting by the chimpanzees of the Nimba Mountains, Guinea: Environmentally or socially determined?" *American Journal of Primatology*, 65: 1–13.

Koops, K., W. C. McGrew, T. Matsuzawa, and L. A. Knapp. 2012. "Terrestrial nest-building by wild chimpanzees (*Pan troglodytes troglodytes*): Implications for the tree-to-ground sleep transition in early hominins." *American Journal of Physical Anthropology*, 148: 351–361.

Koops, K., C. Schöning, M. Isaji, and C. Hashimoto. 2015. "Cultural differences in ant-dipping tool length between neighboring chimpanzee communities at Kalinzu, Uganda." *Scientific Reports*, 5: 12456.

Koops, K., C. Schöning, W. McGrew, and T. Matsuzawa. 2015. "Chimpanzees prey on army ants at Seringbara, Nimba Mountains, Guinea: Predation patterns and tool use characteristics." *American Journal of Primatology*, 77: 319–329.

Kortlandt, A. 1986. "The use of stone tools by wild-living chimpanzees and earliest hominids." *Journal of Human Evolution*, 15: 77–132.

Kret, M. E., and M. Tomonaga. 2016. "Getting to the bottom of face processing: Species-specific inversion effects for faces and behinds in humans and chimpanzees (*Pan troglodytes*)." *PLoS ONE*, 11 (11): e0165357.

Krief, S., M. A. Huffman, T. Sévenet, C. M. Hladik, P. Grellier, P. M. Loiseau, and R. W. Wrangham. 2006. "Bioactive properties of plant species ingested by chimpanzees (*Pan troglodytes schweinfurthii*) in the Kibale National Park, Uganda." *American Journal of Primatology*, 68: 51–71.

Kühl, H. S., A. K. Kalan, M. Arandjelovic, M. Aubert, L. D'Auvergne, A. Goedmakers, S. Jones, et al. 2016. "Chimpanzee accumulative stone throwing." *Scientific Reports*, 6: 22219.

Kühl, H. S., A. N'Guessan, J. Riedel, S. Metzger, and T. Deschner. 2012. "The effect of climate fluctuation on chimpanzee birth sex ratio." *PLoS ONE*, 7 (4): e35610.

Kutsukake, N., M. Teramoto, S. Homma, Y. Mori, K. Matsudaira, H. Kobayashi, T. Ishida, K. Okanoya, and T. Hasegawa. 2011. "Individual variation in behavioural reactions to unfamiliar conspecific vocalization and hormonal underpinnings in male chimpanzees." *Ethology*, 18: 269–280.

Langergraber, K. E., C. Boesch, E. Inoue, M. Inoue-Murayama, J. C. Mitani, T. Nishida, A. Pusey, et al. 2010. "Genetic and 'cultural' similarity in wild chimpanzees." *Proceedings of the Royal Society B (Biological Sciences)*, 278: 408–416.

Langergraber, K. E., J. C. Mitani, and L. Vigilant. 2007. "The limited impact of kinship on cooperation in wild chimpanzees." *Proceedings of the National Academy of Sciences*, 104: 7786–7790.

———. 2009. "Kinship and social bonds in female chimpanzees (*Pan troglodytes*)." *American Journal of Primatology*, 71: 840–851.

Langergraber, K. E., J. C. Mitani, D. P. Watts, and L. Vigilant. 2013. "Male-female socio-spatial relationships and reproduction in wild chimpanzee." *Behavioral Ecology and Sociobiology,* 67: 861–873.

Langergraber, K. E., K. Prüfer, C. Rowney, C. Boesch, C. Crockford, K. Fawcett, E. Inoue, et al. 2012. "Generation times in wild chimpanzees and gorillas suggest earlier divergence times in great ape and human evolution." *Proceedings of the National Academy of Sciences,* 109: 15716–15721.

Langergraber, K. E., C. Rowney, C. Crockford, R. Wittig, K. Zuberbühler, and L. Vigilant. 2014. "Genetic analyses suggest no immigration of adult females and their offspring into the Sonso community of chimpanzees in the Budongo Forest Reserve, Uganda." *American Journal of Primatology,* 76: 640–648.

Langergraber, K. E., C. Rowney, G. Schubert, C. Crockford, C. Hobaiter, R. Wittig, R. W. Wrangham, K. Zuberbühler, and L. Vigilant. 2014. "How old are chimpanzee communities? Time to most recent common ancestor of the Y-chromosome in highly patrilocal societies." *Journal of Human Evolution,* 69: 1–7.

Langergraber, K. E., and L. Vigilant. 2011. "Genetic differences cannot be excluded from generating behavioural differences among chimpanzee groups." *Proceedings of the Royal Society B (Biological Sciences),* 278: 2094–2095.

Laporte, M. N. C., and K. Zuberbühler. 2010. "Vocal greeting behaviour in wild chimpanzee females." *Animal Behaviour,* 80: 467–473.

Launhardt, K., C. Borries, C. Hardt, J. T. Epplen, and P. Winkler. 2001. "Paternity analysis of alternative male reproductive routes among the langurs (*Semnopithecus entellus*) of Ramnagar." *Animal Behaviour,* 61: 53–64.

Lehmann, J., and C. Boesch. 2004. "To fission or to fusion: Effects of community size on wild chimpanzee (*Pan troglodytes verus*) social organization." *Behavioral Ecology and Sociobiology,* 56: 207–216.

———. 2008. "Sexual differences in chimpanzee sociality." *International Journal of Primatology,* 29: 65–81.

Lehmann, J., G. Fickenscher, and C. Boesch. 2007. "Kin biased investment in wild chimpanzees." *Behaviour,* 143: 931–955.

Lind, J., and P. Lindenfors. 2010. "The number of cultural traits is correlated with female group size but not with male group size in chimpanzee communities." *PLoS ONE,* 5 (3): e9241.

Llorente, M., D. Riba, L. Palou, L. Carrasco, M. Mosquera, M. Colell, and O. Feliu. 2011. "Population-level right-handedness for a coordinated bimanual task in naturalistic housed chimpanzees: Replication and extension in 114 animals from Zambia and Spain." *American Journal of Primatology,* 73: 281–290.

Locatelli, S., R. J. Harrigan, P. R. S. Clee, M. W. Mitchell, K. A. Mckean, T. B. Smith, and M. K. Gonder. 2016. "Why are Nigeria-Cameroon chimpanzees (*Pan troglodytes ellioti*) free of SIVcpz infection?" *PLoS ONE*, 11 (8): e0160788.

Locatelli, S., K. A. McKean, P. R. Sesink Clee, and M. K. Gonder. 2014. "The evolution of resistance to simian immunodeficiency virus (SIV): A review." *International Journal of Primatology*, 35: 349–375.

Lonsdorf, E. V. 2005. "Sex differences in the development of termite-fishing skills in the wild chimpanzees, *Pan troglodytes schweinfurthii*, of Gombe National Park, Tanzania." *Animal Behaviour*, 70: 673–683.

Lonsdorf, E. V., K. E. Anderson, M. A. Stanton, M. Shender, M. R. Heintz, J. Goodall, and C. M. Murray. 2014. "Boys will be boys: Sex differences in wild infant chimpanzee social interactions." *Animal Behaviour*, 88: 79–83.

Lonsdorf, E. V., L. E. Eberly, and A. E. Pusey. 2004. "Sex differences in learning in chimpanzees." *Nature*, 428: 715–716.

Lonsdorf, E. V., and W. D. Hopkins. 2005. "Wild chimpanzees show population-level handedness for tool use." *Proceedings of the National Academy of Sciences*, 102: 12634–12638.

Lonsdorf, E. V., A. C. Markham, M. R. Heintz, K. E. Anderson, D. J. Ciuk, J. Goodall, and C. M. Murray. 2014. "Sex differences in wild chimpanzee behavior emerge during infancy." *PLoS ONE*, 9 (6): e99099.

Loudon, J. E., P. A. Sandberg, R. W. Wrangham, B. Fahey, and M. Sponheimer. 2016. "The stable isotope ecology of Pan in Uganda and beyond." *American Journal of Primatology*, 78: 1070–1085.

Lovejoy, C. O. 1978. "A biomechanical review of the locomotor diversity of early hominids." In *Early Hominids of Africa*, edited by C. J. Jolly, 403–429. New York: St. Martin's.

———. 2009. "Reexamining human origins in light of *Ardipithecus ramidus*." *Science*, 326: 74.

Lukas, D., and E. Huchard. 2014. "The evolution of infanticide by males in mammalian societies." *Science*, 346: 841–844.

Luncz, L. V., and C. Boesch. 2014. "Tradition over trend: Neighboring chimpanzee communities maintain differences in cultural behavior despite frequent immigration of adult females." *American Journal of Primatology*, 76: 649–657.

———. 2015. "The extent of cultural variation between adjacent chimpanzee (*Pan troglodytes verus*) communities: A microecological approach." *American Journal of Physical Anthropology*, 156: 67–75.

Lycett, S. J., M. Collard, and W. C. McGrew. 2010. "Are behavioral differences among wild chimpanzee communities genetic or cultural? An assessment using tool-use data and phylogenetic methods." *American Journal of Physical Anthropology*, 142: 461–467.

———. 2011. "Correlations between genetic and behavioural dissimilarities in wild chimpanzees (*Pan troglodytes*) do not undermine the case for culture." *Proceedings of the Royal Society B (Biological Sciences),* 278: 2091–2093.

Machanda, Z. P., I. C. Gilby, and R. W. Wrangham. 2014. "Mutual grooming among adult male chimpanzees: The immediate investment hypothesis." *Animal Behaviour,* 87: 165–174.

Makanga, B., P. Yangari, N. Rahola, V. Rougeron, E. Elguero, L. Boundenga, N. D. Moukodoum, et al. 2016. "Ape malaria transmission and potential for ape-to-human transfers in Africa." *Proceedings of the National Academy of Sciences,* 113: 5329–5334.

Manson, J. H., and R. W. Wrangham. 1991. "Intergroup aggression in chimpanzees and humans." *Current Anthropology,* 32: 369–390.

Markham, A. C., E. V. Lonsdorf, A. E. Pusey, and C. M. Murray. 2015. "Maternal rank influences the outcome of aggressive interactions between immature chimpanzees." *Animal Behaviour,* 100: 192–198.

Matsumoto, T., N. Itoh, S. Inoue, and M. Nakamura. 2016. "An observation of a severely disabled infant chimpanzee in the wild and her interactions with her mother." *Primates,* 57: 3–7.

Matsumoto-Oda, A. 1999. "Female choice in the opportunistic mating of wild chimpanzees (*Pan troglodytes schweinfurthii*) at Mahale." *Behavioral Ecology and Sociobiology,* 46: 258–266.

Matsumoto-Oda, A., M. Hamai, H. Hayaki, K. Hosaka, K. D. Hunt, E. Kasuya, K. Kawanaka, J. C. Mitani, H. Takasaki, and Y. Takahata. 2007. "Estrus cycle asynchrony in wild female chimpanzees, *Pan troglodytes schweinfurthii.*" *Behavioral Ecology and Sociobiology,* 61: 661–668.

Matsumoto-Oda, A., and Y. Hayashi. 1999. "Nutritional aspects of fruit choice by chimpanzees." *Folia Primatologica,* 70: 154–162.

Matsumoto-Oda, A., K. Hosaka, M. A. Huffman, and K. Kawanaka. 1998. "Factors affecting party size in chimpanzees of the Mahale Mountains." *International Journal of Primatology,* 19: 999–1011.

Matsumoto-Oda, A., and Y. Ihara. 2011. "Estrous asynchrony causes low birth rates in wild female chimpanzees." *American Journal of Primatology,* 73: 180–188.

Matsuzawa, T. 2006. "Sociocognitive development in chimpanzees: A synthesis of laboratory work and fieldwork." In *Cognitive Development in Chimpanzees,* edited by T. Matsuzawa, M. Tomonaga, and M. Tanaka, 3–33. Tokyo: Springer.

McBrearty, S., and N. G. Jablonski. 2005. "First fossil chimpanzee." *Nature,* 437: 105–108.

McCarthy, M. S., C. E. Finch, and C. B. Stanford. 2013. "Alpha male status predicts longevity in wild male chimpanzees." *American Journal of Physical Anthropology,* 150: 193 (abstract).

McCarthy, M. S., J. D. Lester, E. J. Howe, M. Arandelovic, C. B. Stanford, and L. Vigilant. 2015. "Genetic censusing identifies an unexpectedly sizeable population of an endangered large mammal in a fragmented forest landscape." *BMC Ecology,* 15: 21.

McCarthy, M. S., J. D. Lester, and C. B. Stanford. 2017. "Chimpanzees (*Pan troglodytes*) flexibly use introduced species for nesting and bark feeding in a human-dominated habitat." *International Journal of Primatology,* 38: 321–337.

McGrew, W. C. 1992. *Chimpanzee Material Culture.* Cambridge: Cambridge University Press.

———. 2004. *The Cultured Chimpanzee.* Cambridge: Cambridge University Press.

———. 2014. "The 'other faunivory' revisited: Insectivory in human and non-human primates and the evolution of human diet." *Journal of Human Evolution,* 71: 4–11.

———. 2015. "Snakes as hazards: Modeling risk by chasing chimpanzees." *Primates,* 56: 107–111.

———. 2017. "Field studies of *Pan troglodytes* reviewed and comprehensively mapped, focussing on Japan's contribution to cultural primatology." *Primates,* 58: 237–258.

McGrew, W. C., P. J. Baldwin, and C. E. G. Tutin. 1981. "Chimpanzees in a hot, dry and open habitat: Mt. Assirik, Senegal, West Africa." *Journal of Human Evolution,* 10: 227–244.

McGrew, W. C., and L. F. Marchant. 1997. "On the other hand: Current issues in and meta-analysis of the behavioral laterality of hand function in nonhuman primates." *Yearbook of Physical Anthropology,* 40: 201–232.

McGrew, W. C., L. F. Marchant, and T. Nishida. 1996. *Great Ape Societies.* Cambridge: Cambridge University Press.

McLennan, M. R. 2011. "Tool-use to obtain honey by chimpanzees at Bulindi: New record from Uganda." *Primates,* 52: 315–322.

———. 2014. "Chimpanzee insectivory in the northern half of Uganda's Rift Valley: Do Bulindi chimpanzees conform to a regional pattern?" *Primates,* 55: 173–178.

———. 2015. "Is honey a fallback food for wild chimpanzees or just a sweet treat?" *American Journal of Physical Anthropology,* 158: 685–695.

Mech, L. D. 1994. "Buffer zones of territories of gray wolves as regions of intraspecific strife." *Journal of Mammalogy,* 75: 199–202.

Mech, L. D., D. W. Smith, K. M. Murphy, and D. R. MacNulty. 2001. "Winter severity and wolf predation on a formerly wolf-free elk herd." *Journal of Wildlife Management,* 65: 998–1003.

Mercader, J., H. Barton, J. Gillespie, J. Harris, S. Kuhn, R. Tyler, and C. Boesch. 2007. "4,300-year-old chimpanzee sites and the origins of percussive stone technology." *Proceedings of the National Academy of Sciences,* 104: 3043–3048.

Mercader, J., M. Panger, and C. Boesch. 2002. "Excavation of a chimpanzee stone tool site in the African rainforest." *Science,* 296: 1452–1455.

Mikkelsen, T. S., L. W. Hillier, E. E. Eichler, M. C. Zody, D. B. Jaffe, S. P. Yang, W. Enard et al. 2005. "Initial sequence of the chimpanzee genome and comparison with the human genome." *Nature,* 437: 69–87.

Miller, J. A., A. E. Pusey, I. C. Gilby, K. Schroepfer-Walker, A. C. Markham, and C. M. Murray. 2014. "Competing for space: Female chimpanzees are more aggressive inside than outside their core areas." *Animal Behaviour,* 87: 147–152.

Mills, M. G. L., L. S. Broomhall, and J. T. du Toit. 2004. "Cheetah *Acinonyx jubatus* feeding ecology in the Kruger National Park and a comparison across African savanna habitats: Is the cheetah only a successful hunter on open grassland plains?" *Wildlife Biology,* 10: 177–186.

Mitani, J. C. 2009. "Male chimpanzees form enduring and equitable social bonds." *Animal Behaviour,* 77: 633–640.

Mitani, J. C., K. L. Hundley, and M. E. Murdoch. 1999. "Geographic variation in the calls of wild chimpanzees: A reassessment." *American Journal of Primatology,* 47: 133–151.

Mitani, J. C., D. P. Watts, and S. J. Amsler. 2010. "Lethal intergroup aggression leads to territorial expansion in wild chimpanzees." *Current Biology,* 20: R507–R508.

Mitani, J. C., D. P. Watts, and M. N. Muller. 2002. "Recent developments in the study of wild chimpanzee behavior." *Evolutionary Anthropology,* 11: 9–25.

Moore, D. L., K. E. Langergraber, and L. Vigilant. 2015. "Genetic analyses suggest male philopatry and territoriality in savanna-woodland chimpanzees (*Pan troglodytes schweinfurthii*) of Ugalla, Tanzania." *International Journal of Primatology,* 36: 377–397.

Moore, J. 1996. "Savanna chimpanzees, referential models and the last common ancestor." In *Great Ape Societies,* edited by W. C. McGrew, L. F. Marchant, and T. Nishida, 275–292. Cambridge: Cambridge University Press.

Moorjani, P., C. E. Amorim, P. F. Arndt, and M. Przeworski. 2016. "Variation in the molecular clock of primates." *Proceedings of the National Academy of Sciences,* 113: 10607–10612.

Morgan, D., and C. Sanz. 2006. "Chimpanzee feeding ecology and comparisons with sympatric gorillas in the Goualougo Triangle, Republic of Congo." In *Feeding Ecology in Apes and Other Primates,* edited by G. Hohmann, M. M. Robbins, and C. Boesch, 97–122. Cambridge: Cambridge University Press.

Muehlenbein, M. P., and D. P. Watts. 2010. "The costs of dominance: Testosterone, cortisol and intestinal parasites in wild male chimpanzees." *BioPscyhoSocial Medicine,* 4: 21–22.

Muehlenbein, M. P., D. P. Watts, and P. L. Whitten. 2004. "Dominance rank and fecal testosterone levels in adult male chimpanzees (*Pan troglodytes schweinfurthii*) at Ngogo, Kibale National Park, Uganda." *American Journal of Primatology,* 64: 71–82.

Muller, M. N. 2007. "Chimpanzee violence: Femmes fatales." *Current Biology,* 17: R355–R356.

Muller, M. N., M. Emery Thompson, S. M. Kahlenberg, and R.W. Wrangham. 2011. "Sexual coercion by male chimpanzees shows that female choice may be more apparent than real." *Behavioral Ecology and Sociobiology,* 65: 921–933.

Muller, M. N., M. Emery Thompson, and R. W. Wrangham. 2006. "Male chimpanzees prefer mating with old females." *Current Biology,* 16: 2234–2238.

Muller, M. N., S. M. Kahlenberg, M. Emery Thompson, and R. W. Wrangham. 2007. "Male coercion and the costs of promiscuous mating for female chimpanzees." *Proceedings of the Royal Society B (Biological Sciences),* 274: 1009–1014.

Muller, M. N., S. M. Kahlenberg, and R. W. Wrangham. 2009. "Male aggression against females and sexual coercion in chimpanzees." In *Sexual Coercion in Primates and Humans: An Evolutionary Perspective on Male Aggression against Females,* edited by M. N. Muller and R. W. Wrangham, 184–217. Cambridge, Mass.: Harvard University Press.

Muller, M. N., and S. F. Lipson. 2003. "Diurnal patterns of urinary steroid excretion in wild chimpanzees." *American Journal of Primatology,* 60: 161–166.

Muller, M. N., and R. W. Wrangham. 2004a. "Dominance, aggression and testosterone in wild chimpanzees: A test of the 'challenge hypothesis.' " *Animal Behaviour,* 67: 113–123.

———. 2004b. "Dominance, cortisol and stress in wild chimpanzees (*Pan troglodytes schweinfurthii*)." *Behavioral Ecology and Sociobiology,* 55: 332–340.

———. 2014. "Mortality rates among Kanyawara chimpanzees." *Journal of Human Evolution,* 66: 107–114.

Murray, C. M., I. C. Gilby, S. V. Mane, and A. E. Pusey. 2008. "Adult male chimpanzees inherit maternal ranging patterns." *Current Biology,* 18: 20–24.

Murray, C. M., E. V. Lonsdorf, M. A. Stanton, K. R. Wellens, J. A. Miller, J. Goodall, and A. E. Pusey. 2014. "Early social exposure in wild chimpanzees: Mothers with sons are more gregarious than mothers with daughters." *Proceedings of the National Academy of Sciences,* 111: 18189–18194.

Murray, C. M., S. V. Mane, and A. E. Pusey. 2007. "Dominance rank influences female space use in wild chimpanzees, *Pan troglodytes:* Towards an ideal despotic distribution." *Animal Behaviour,* 74: 1795–1804.

Murray, C. M., E. E. Wroblewski, and A. E. Pusey. 2007. "New case of intragroup infanticide in the chimpanzees of Gombe National Park." *International Journal of Primatology*, 28: 23–37.

Musgrave, S., D. Morgan, E. Lonsdorf, R. Mundry, and C. Sanz. 2016. "Tool transfers are a form of teaching among chimpanzees." *Scientific Reports*, 6: 34783.

Nakamura, M., H. Hayaki, K. Hosaka, N. Itoh, and K. Zamma. 2014. "Brief communication: Orphaned male chimpanzees die young even after weaning." *American Journal of Physical Anthropology*, 153: 139–143.

Nakamura, M., and K. Hosaka. 2015. "Orphans and allomothering." In *Mahale Chimpanzees: 50 Years of Research*, edited by M. Nakamura, K. Hosaka, N. Itoh, and K. Zamma, 421–432. Cambridge: Cambridge University Press.

Nakamura, M., K. Hosaka, N. Itoh, and K. Zamma, eds. 2015. *Mahale Chimpanzees: 50 Years of Research*. Cambridge: Cambridge University Press.

Nakamura, M., and N. Itoh. 2015. "Conspecific killings." In *Mahale Chimpanzees: 50 Years of Research*, edited by M. Nakamura, K. Hosaka, N. Itoh, and K. Zamma, 372–383. Cambridge: Cambridge University Press.

Newton-Fisher, N. E. 1999. "The diet of chimpanzees in the Budongo Forest." *African Journal of Ecology*, 34: 344–354.

———. 2004. "Hierarchy and social status in Budongo chimpanzees." *Primates*, 45: 81–87.

———. 2006. "Female coalitions against male aggression in wild chimpanzees of the Budongo Forest." *International Journal of Primatology*, 27: 1589–1599.

———. 2014. "Roving females and patient males: A new perspective on the mating strategies of chimpanzees." *Biological Reviews*, 89: 356–374.

Newton-Fisher, N. E., V. Reynolds, and A. J. Plumptre. 2000. "Food supply and chimpanzee (*Pan troglodytes schweinfurthii*) party size in the Budongo Forest Reserve, Uganda." *International Journal of Primatology*, 21: 613–628.

Newton-Fisher, N. E., M. Emery Thompson, V. Reynolds, C. Boesch, and L. Vigilant. 2010. "Paternity and social rank in wild chimpanzees (*Pan troglodytes*) from the Budongo Forest, Uganda." *American Journal of Physical Anthropology*, 142: 417–428.

N'guessan, A. K., S. Ortmann, and C. Boesch. 2009. "Daily energy balance and protein gain among *Pan troglodytes verus* in the Taï National Park, Cote d'Ivoire." *International Journal of Primatology*, 30: 481–496.

Nishida, T. 1968. "The social group of wild chimpanzees in the Mahali Mountains." *Primates*, 9: 167–224.

———. 1983a. "Alloparental behavior in wild chimpanzees of the Mahale Mountains, Tanzania." *Folia Primatologica*, 41: 1–33.

———. 1983b. "Alpha status and agonistic alliance in wild chimpanzees (*Pan troglodytes schweinfurthii*)." *Primates*, 24: 318–336.

———. 1990. *The Chimpanzees of the Mahale Mountains*. Tokyo: Tokyo University Press.

———. 1996. "The death of Ntologi, the unparalleled leader of M-group." *Pan Africa News*, 3: 1–4.

———. 1997. "Sexual behavior of adult male chimpanzees of the Mahale Mountains National Park, Tanzania." *Primates*, 38: 379–398.

Nishida, T., N. Corp, M. Hamai, T. Hasegawa, M. Hiraiwa-Hasegwawa, K. Hoskaa, K. D. Hunt, et al. 2003. "Demography, female life history, and reproductive profiles among the chimpanzees of Mahale." *American Journal of Primatology*, 59: 99–121.

Nishida, T., T. Hasegawa, H. Hayaki, Y. Takahata, and S. Uehara. 1992. "Meat-sharing as a coalition strategy by an alpha male chimpanzee?" In *Topics in Primatology*, vol. 1, *Human Origins*, edited by T. Nishida, W. C. McGrew, P. Marler, M. Pickford, and F. de Waal, 159–174. Tokyo: Tokyo University Press.

Nishida, T., T. Matsusaka, and W. C. McGrew. 2009. "Emergence, propagation or disappearance of novel behavioral patterns in the habituated chimpanzees of Mahale: A review." *Primates*, 50: 23–36.

Nishida, T., S. Uehara, and R. Nyundo. 1979. "Predatory behavior among wild chimpanzees of the Mahale Mountains." *Primates*, 20: 1–20.

Normand, E., S. D. Ban, and C. Boesch. 2009. "Forest chimpanzees (*Pan troglodytes verus*) remember the location of numerous fruit trees." *Animal Cognition*, 12: 797–807.

Oelze, V. M., G. Fahy, G. Hohmann, M. M. Robbins, V. Leinert, K. Lee, H. Eshuis, et al. 2016. "Comparative isotope ecology of African great apes." *Journal of Human Evolution*, 101: 1–16.

Olsen, M. V., and A. Varki. 2003. "Sequencing the chimpanzee genome: Insights into human evolution and disease." *Nature Reviews Genetics*, 4: 20–28.

Olshansky, S. J. 2011. "Aging of U.S. presidents." *Journal of the American Medical Association*, 306 (21): 2328–2329.

O'Malley, R. C., and M. L. Power. 2012. "Nutritional composition of actual and potential insect prey for the Kasekela chimpanzees of Gombe National Park, Tanzania." *American Journal of Physical Anthropology*, 149: 493–503.

O'Malley, R. C., M. A. Stanton, I. C. Gilby, E. V. Lonsdorf, A. Pusey, A. C. Markham, and C. M. Murray. 2016. "Reproductive state and rank influence patterns of meat consumption in wild female chimpanzees (*Pan troglodytes schweinfurthii*)." *Journal of Human Evolution*, 90: 16–28.

O'Malley, R. C., W. Wallauer, C. M. Murray, and J. Goodall. 2012. "The appearance and spread of ant fishing among the Kasekela chimpanzees." *Current Anthropology*, 53: 650–663.

Oshawa, H., M. Ioue, and O. Takenaka. 1993. "Mating strategy and reproductive success of male patas monkeys (*Erythrocebus patas*)." *Primates*, 34: 533–544.

Otali, E., and J. S. Gilchrist. 2006. "Why chimpanzee (*Pan troglodytes schweinfurthii*) mothers are less gregarious than nonmothers and males: The infant safety hypothesis." *Behavioral Ecology and Sociobiology*, 59: 561–570.

Pascual-Garrido, A., B. Umaru, O. Allon, and V. Sommer. 2013. "Apes finding ants: Predator-prey dynamics in a chimpanzee habitat in Nigeria." *American Journal of Primatology*, 75: 1231–1244.

Patterson, N., D. J. Richter, S. Gnerre, E. S. Lander, and D. Reich. 2006. "Genetic evidence for complex speciation of humans and chimpanzees." *Nature*, 441: 1103–1108.

Pearson, H. C. 2011. "Sociability of female bottlenose dolphins (*Tursiops spp.*) and chimpanzees (*Pan troglodytes*): Understanding evolutionary pathways toward social convergence." *Evolutionary Anthropology*, 20: 85–95.

Pemberton, J. M., S. D. Albon, F. E. Guinness, T. H. Clutton-Brock, and G. A. Dover. 1992. "Behavioral estimates of male mating success tested by DNA fingerprinting in a polygynous mammal." *Behavioral Ecology*, 3: 66–75.

Phillips, C. A., and T. C. O'Connell. 2016. "Fecal carbon and nitrogen isotopic analysis as an indicator of diet in Kanyawara chimpanzees, Kibale National Park, Uganda." *American Journal of Physical Anthropology*, 161: 685–697.

Plooij, F. X. 1984. *The Behavioral Development of Free-Living Chimpanzee Babies and Infants.* Norwood, N.J.: Ablex.

Pontzer, H., and R. W. Wrangham. 2006. "Ontogeny of ranging in wild chimpanzees." *International Journal of Primatology*, 27: 295–309.

Pope, T. 1990. "The reproductive consequences of male cooperation in the red howler monkey: Paternity exclusion in multi-male and single male troops using genetic markers." *Behavioral Ecology and Sociobiology*, 27: 439–446.

Potts, K. B., E. Baken, A. Leaving, and D. P. Watts. 2016. "Ecological factors influencing habitat use by chimpanzees at Ngogo, Kibale National Park, Uganda." *American Journal of Primatology*, 78: 432–440.

Potts, K. B., E. Baken, S. Ortman, D. P. Watts, and R. W. Wrangham. 2015. "Variability in population density is paralleled by large differences in foraging efficiency in chimpanzees." *International Journal of Primatology*, 36: 1101–1119.

Potts, K. B., D. P. Watts, and R. W. Wrangham. 2011. "Comparative feeding ecology of two communities of chimpanzees (*Pan troglodytes*) in Kibale National Park, Uganda." *International Journal of Primatology*, 32: 669–690.

Port, M., and P. M. Kappeler. 2010. "The utility of reproductive skew models in the study of male primates, a critical evaluation." *Evolutionary Anthropology*, 19: 46–56.

Power, M. 1991. *The Egalitarians—Human and Chimpanzee: An Anthropological View of Social Organization.* Cambridge: Cambridge University Press.

Pradhan, G. R., S. A. Pandit, and C. P. Van Schaik. 2014. "Why do chimpanzee males attack the females of neighboring communities?" *American Journal of Physical Anthropology,* 155: 430–435.

Prado-Martinez, J., P. H. Sudmant, J. M. Kidd, H. Li, J. L. Kelley, B. Lorente-Galdos, K. R. Veeramah, et al. 2013. "Great ape genetic diversity and population history." *Nature,* 499: 471–475.

Pruetz, J. D. 2007. "Evidence of cave use by savanna chimpanzees (*Pan troglodytes verus*) at Fongoli, Senegal: Implications for thermoregulatory behavior." *Primates,* 48: 316–319.

Pruetz, J. D., and P. Bertolani. 2007. "Savanna chimpanzees, *Pan troglodytes verus,* hunt with tools." *Current Biology,* 17: 412–417.

———. 2009. "Chimpanzee (*Pan troglodytes verus*) behavioral responses to stress associated with living in a savanna-mosaic environment: Implications for hominin adaptations to open habitats." *PaleoAnthropology,* 2009: 252–262.

Pruetz, J. D., P. Bertolani, K. Boyer Ontl, S. Lindshield, M. Shelley, and E. G. Wessling. 2015. "New evidence on the tool-assisted hunting exhibited by chimpanzees (*Pan troglodytes verus*) in a savannah habitat at Fongoli, Senegal." *Royal Society Open Science,* 2: 140507.

Pruetz, J. D., K. Boyer Ontl, E. Cleaveland, S. Lindshield, J. Marshack, and E. G. Wessling. 2017. "Intragroup lethal aggression in West African chimpanzees (*Pan troglodytes verus*): Inferred killing of a former alpha male at Fongoli, Senegal." *International Journal of Primatology,* 38 (1): 31–57.

Pruetz, J. D., and S. Lindshield. 2012. "Plant-food and tool transfer among savanna chimpanzees at Fongoli, Senegal." *Primates,* 53: 133–145.

Prüfer, K., K. Munch, I. Hellman, K. Akagi, J. R. Miller, B. Walenz, S. Koren, et al. 2015. "The bonobo genome compared with the chimpanzee and human genomes." *Nature,* 486: 527–531.

Pusey, A. E. 1983. "Mother-infant relationships in chimpanzees after weaning." *Animal Behaviour,* 31: 363–377.

———. 1990. "Behavioural changes at adolescence in chimpanzees." *Behaviour,* 115: 203–246.

Pusey, A. E., and C. Packer. 1994. "Infanticide in lions: Consequences and counterstrategies." In *Infanticide and Parental Care,* edited by S. Parmigiani and F. vom Saal, 277–299. London: Harwood Academic.

Pusey, A. E., J. Williams, and J. Goodall. 1997. "The influence of dominance rank on the reproductive success of female chimpanzees." *Science,* 277: 828–831.

Pusey, A. E., M. L. Wilson, and D. A. Collins, 2008. "Human impacts, disease risk and population dynamics in the chimpanzees of Gombe National Park, Tanzania." *American Journal of Primatology,* 70: 738–744.

Reidel, J., M. Franz, and C. Boesch. 2011. "How feeding competition determines female chimpanzee gregariousness and ranging in the Taï National Park, Cote d'Ivoire." *American Journal of Primatology*, 73: 305–313.

Roberts, A. I., and S. G. B. Roberts. 2016. "Wild chimpanzees modify modality of gestures according to strength of social bonds and personal network size." *Scientific Reports*, 6: 33864.

Rogers, A., D. Iltis, and S. Wooding. 2004. "Genetic variation at the MC1R locus and the time since loss of human body hair." *Current Anthropology*, 45: 105–108.

Rowell, T. E. 1974. "The concept of social dominance." *Behavioral Biology*, 11: 131–154.

Ruvolo, M. 1997. "Molecular phylogeny of the hominoids: Inferences from multiple independent DNA sequence data sets." *Molecular Biology and Evolution*, 14: 248–265.

Samson, D. R., and K. D. Hunt. 2012. "A thermodynamic comparison of arboreal and terrestrial sleeping sites for dry-habitat chimpanzees (*Pan troglodytes schweinfurthii*) at the Toro-Semliki Wildlife Reserve, Uganda." *American Journal of Primatology*, 74: 811–818.

———. 2014. "Chimpanzees preferentially select sleeping platform construction tree species with biomechanical properties that yield stable, firm, but compliant nests." *PLoS ONE*, 9 (4): e95361.

Samuni, L., A. Preis, R. Mundry, T. Deschner, C. Crockford, and R. M. Wittig. 2017. "Oxytocin reactivity during intergroup conflict in wild chimpanzees." *Proceedings of the National Academy of Sciences*, 114: 268–273.

Sand, H., C. Wikenros, P. Wabakken, and O. Liberg. 2006. "Effects of hunting group size, snow depth and age on the success of wolves hunting moose." *Animal Behaviour*, 72: 781–789.

Sandel, A. A., R. B. Reddy, and J. C. Mitani. 2017. "Adolescent male chimpanzees do not form a dominance hierarchy with their peers." *Primates*, 58: 39–49.

Sanz, C. M., and D. B. Morgan. 2009. "Flexible and persistent tool-using strategies in honey gathering by wild chimpanzees." *International Journal of Primatology*, 30: 411–427.

Sanz, C. M., D. B. Morgan, and S. Gulick. 2004. "New insights into chimpanzees, tools, and termites from the Congo basin." *American Naturalist*, 164: 567–581.

Sanz, C. M., D. B. Morgan, and W. D. Hopkins. 2016. "Lateralization and performance asymmetries in the termite fishing of wild chimpanzees in the Goualougo Triangle, Republic of Congo." *American Journal of Primatology*, 78: 1190–1200.

Sanz, C. M., C. Schöning, and D. B. Morgan. 2010. "Chimpanzees prey on army ants with specialized tool set." *American Journal of Primatology*, 72: 17–24.

Sapolsky, R. M. 1992. "Cortisol concentrations and the social significance of rank instability among wild baboons." *Psychoneuroendocrinology*, 17: 701–709.

———. 2005. "The influence of social hierarchy on primate health." *Science*, 308: 648–652.

Sarich, V. M., and A. C. Wilson. 1967. "Immunological time scale for hominid evolution." *Science*, 158: 1200–1203.

Sarmiento, E., T. Butynski, and J. Kalina. 1996. "Ecological, morphological, and behavioral aspects of gorillas of Bwindi-Impenetrable and Virungas National Parks, with implications for gorilla taxonomic affinities." *American Journal of Primatology*, 40: 1–21.

Sayers, K., and C. O. Lovejoy. 2008. "The chimpanzee has no clothes: A critical examination of *Pan troglodytes* in models of modern evolution." *Current Anthropology*, 49: 87–114.

Schel, A. M., Z. Machanda, S. W. Townsend, K. Zuberbühler, and K. E. Slocombe. 2013. "Chimpanzee food calls are directed at specific individuals." *Animal Behaviour*, 86: 955–965.

Schoeninger, M. J., J. Moore, and J. M. Sept. 1999. "Subsistence strategies of two 'savanna' chimpanzee populations: The stable isotope evidence." *American Journal of Primatology*, 49: 297–314.

Schoeninger, M. J., C. A. Most, J. J. Moore, and A. D. Somerville. 2016. "Environmental variables across *Pan troglodytes* study sites correspond with the carbon, but not the nitrogen, stable isotope ratios of chimpanzee hair." *American Journal of Primatology*, 78: 1055–1069.

Schöning, C., D. Ellis, A. Fowler, and V. Sommer. 2007. "Army ant prey availability and consumption by chimpanzees at Gashaka (Nigeria)." *Journal of Zoology*, 271: 125–133.

Schöning, C., T. Humle, Y. Möbius, and W. C. McGrew. 2008. "The nature of culture: Technological variation in chimpanzee predation on army ants revisited." *Journal of Human Evolution*, 55: 48–59.

Schrauf, C., J. Call, K. Fuwa, and S. Hirata. 2012. "Do chimpanzees use weight to select hammer tools?" *PLoS ONE*, 7 (7): e41044.

Schubert, G., L. Vigilant, C. Boesch, R. Klenke, K. Langergraber, R. Mundry, M. Surbeck, and G. Hohmann. 2013. "Co-residence between males and their mothers and grandmothers is more frequent in bonobos than chimpanzees." *PLoS ONE*, 8 (12): e83870.

Schülke, O., and J. Ostner. 2012. "Ecological and social influences on sociality." In *The Evolution of Primate Societies*, edited by J. C. Mitani, J. Call, P. M. Kappeler, R. A. Palombit, and J. B. Silk, 195–219. Chicago: University of Chicago Press.

Sherrow, H. 2012. "Adolescent male chimpanzees at Ngogo, Kibale National Park, Uganda have decided dominance relationships." *Folia Primatologica*, 83: 67–75.

Shimada, M. 2013. "Dynamics of the temporal structures of playing clusters and cliques among wild chimpanzees in Mahale Mountains National Park." *Primates,* 54: 245–257.

Sibley, C. G., and J. E. Ahlquist. 1984. "The phylogeny of the hominoid primates, as indicated by DNA–DNA hybridization." *Journal of Molecular Evolution,* 20: 2–15.

Silk, J. B. 2014. "The evolutionary roots of lethal conflict." *Nature,* 513: 321–322.

Sirianni, G., R. Mundry, and C. Boesch. 2015. "When to choose which tool: Multidimensional and conditional selection of nut-cracking hammers in wild chimpanzees." *Animal Behaviour,* 100: 152–165.

Slocombe, K. E., T. Kaller, L. Turman, S. W. Townsend, S. Papworth, P. Squibbs, and K. Zuberbühler. 2010. "Production of food-associated calls in wild male chimpanzees is dependent on the composition of the audience." *Behavioral Ecology and Sociobiology,* 64: 1959–1966.

Smith, T. M., Z. Machanda, A. B. Bernard, R. M. Donovan, A. M. Papakyrikos, M. N. Muller, and R. W. Wrangham. 2013. "First molar eruption, weaning, and life history in living wild chimpanzees." *Proceedings of the National Academy of Sciences,* 110: 2787–2791.

Smuts, B. B. 1985. *Sex and Friendship in Baboons.* New York: Aldine.

Sobolewski, M. E., J. L. Brown, and J. C. Mitani. 2012. "Territoriality, tolerance and testosterone in wild chimpanzees." *Animal Behaviour,* 84: 1469–1474.

Sommer, V., U. Buba, G. Jesus, and A. Pascual-Garrido. 2012. "Till the last drop: Honey gathering in Nigerian chimpanzees." *Ecotropica,* 18: 55–64.

Speth, J. D., and D. D. Davis. 1976. "Seasonal variability in early hominid predation." *Science,* 192: 441–445.

Stanford, C. B. 1995. "The influence of chimpanzee predation on group size and anti-predator behaviour in red colobus monkeys." *Animal Behaviour,* 49: 577–587.

———. 1996. "The hunting ecology of wild chimpanzees: Implications for the behavioral ecology of Pliocene hominids." *American Anthropologist,* 98: 96–113.

———. 1998. *Chimpanzee and Red Colobus: The Ecology of Predator and Prey.* Cambridge, Mass.: Harvard University Press.

———. 1999. *The Hunting Apes: Meat-Eating and the Origins of Human Behavior.* Princeton, N.J.: Princeton University Press.

———. 2001. "The subspecies concept in primatology: The case of mountain gorillas." *Primates,* 42: 309–318.

———. 2006. "Arboreal bipedalism in wild chimpanzees: Implications for the evolution of hominid posture and locomotion." *American Journal of Physical Anthropology,* 129: 225–231.

———. 2012a. "Chimpanzees and the behavior of *Ardipithecus ramidus.*" *Annual Reviews in Anthropology,* 41: 139–149.

———. 2012b. *Planet without Apes*. Cambridge, Mass.: Harvard University Press.
Stanford, C. B., C. Gamboneza, J. B. Nkurunungi, and M. L. Goldsmith. 2000. "Chimpanzees in Bwindi-Impenetrable National Park, Uganda, use different tools to obtain different types of honey." *Primates*, 41: 337–341.
Stanford, C. B., and J. B. Nkurunungi. 2003. "Behavior ecology of sympatric chimpanzees and gorillas in Bwindi Impenetrable National Park, Uganda: Diet." *International Journal of Primatology*, 24: 901–918.
Stanford, C. B., and R. C. O'Malley. 2008. "Sleeping tree choice by Bwindi chimpanzees." *American Journal of Primatology*, 70: 642–649.
Stanford, C. B., J. Wallis, H. Matama, and J. Goodall. 1994a. "Patterns of predation by chimpanzees on red colobus monkeys in Gombe National Park, Tanzania, 1982–1991." *American Journal of Physical Anthropology*, 94: 213–228.
Stanford, C. B., J. Wallis, E. Mpongo, and J. Goodall. 1994b. "Hunting decisions in wild chimpanzees." *Behaviour*, 131: 1–20.
Stanton, M. A., E. V. Lonsdorf, A. E. Pusey, J. Goodall, and C. M. Murray. 2014. "Maternal behavior by birth order in wild chimpanzees (*Pan troglodytes*)." *Current Anthropology*, 55: 483–489.
Stern, J. T., and R. L. Susman. 1983. "The locomotor anatomy of *Australopithecus afarensis*." *American Journal of Physical Anthropology*, 60: 279–317.
Stewart, F. A. 2011. "Why sleep in a nest? Empirical testing of the function of simple shelters made by wild chimpanzees." *American Journal of Physical Anthropology*, 146: 313–318.
Stewart, F. A., and A. K. Piel. 2014. "Termite fishing by wild chimpanzees: New data from Ugalla, western Tanzania." *Primates*, 55: 35–40.
Stewart, F. A., and J. D. Pruetz. 2013. "Do chimpanzee nests serve an antipredatory function?" *American Journal of Primatology*, 75: 593–604.
Strum, S. C. 1982. "Agonistic dominance in male baboons: An alternative view." *International Journal of Primatology*, 3: 175–202.
Stumpf, R. M., and C. Boesch. 2005. "Does promiscuous mating preclude female choice? Female sexual strategies in chimpanzees (*Pan troglodytes verus*) of the Taï National Park, Cote d'Ivoire." *Behavioral Ecology and Sociobiology*, 57: 511–524.
———. 2006. "The efficacy of female choice in chimpanzees of the Taï Forest, Cote d'Ivoire." *Behavioral Ecology and Sociobiology*, 60: 749–765.
———. 2010. "Male aggression and sexual coercion in wild West African chimpanzees, *Pan troglodytes verus*." *Animal Behaviour*, 79: 333–342.
Stumpf, R. M., M. Emery Thompson, M. N. Muller, and R. W. Wrangham. 2009. "The context of female dispersal in Kanyawara chimpanzees." *Behaviour*, 146: 629–656.

Sugiyama, Y., S. Kawamoto, O. Takenaka, K. Kumazaki, and N. Miwa. 1993. "Paternity discrimination and inter-group relationships of chimpanzees at Bossou." *Primates,* 34: 545–552.

Sugiyama, Y., and J. Koman. 1979. "Social structure and dynamics of wild chimpanzees at Bossou, Guinea." *Primates,* 20: 513–524.

Susman, R. L., J. T. Stern, and W. L. Jungers. 1984. "Arboreality and bipedality in the Hadar hominids." *Folia Primatologica,* 43: 113–156.

Sussman, R. W. and J. L. Marshack 2013. "Are humans inherently killers?" Global Nonkilling Working Papers 1, Center for Global Nonkilling, Honolulu.

Tagg, N., J. Willie, C. A. Petre, and O. Haggis. 2013. "Ground night nesting in chimpanzees: New insights from central chimpanzees (*Pan troglodytes troglodytes*) in South-East Cameroon." *Folia Primatologica,* 84: 362–383.

Takahata, Y. 1990. "Adult males' social relations with adult females." In *The Chimpanzees of the Mahale Mountains: Sexual and Life-History Strategies,* edited by T. Nishida, 149–170. Tokyo: University of Tokyo Press.

———. 2015. "Disappearance of K-group male chimpanzees: Re-examination of group extinction." In *Mahale Chimpanzees: 50 Years of Research,* edited by M. Nakamura, K. Hosaka, N. Itoh, and K. Zamma, 119–127. Cambridge: Cambridge University Press.

Takahata, Y., T. Hasegawa, and T. Nishida. 1984. "Chimpanzee predation in the Mahale Mountains from August 1979 to May 1982." *International Journal of Primatology,* 5: 213–233.

Takemoto, H., Y. Kawamoto, and T. Furuichi. 2015. "How did bonobos come to range south of the Congo River? Reconsideration of the divergence of *Pan paniscus* from other *Pan* populations." *Evolutionary Anthropology,* 24: 170–184.

Teleki, G. 1973. *The Predatory Behavior of Wild Chimpanzees.* Lewisburg, Pa.: Bucknell University Press.

Tennie, C., R. O. O'Malley, and I. C. Gilby. 2014. "Why do chimpanzees hunt? Considering the benefits and costs of acquiring and consuming vertebrate versus invertebrate prey." *Journal of Human Evolution,* 71: 38–45.

Terio, K. A., M. J. Kinsel, J. Raphael, T. Mlengeya, I. Lipende, C. A. Kirchoff, B. Gilagiza, et al. 2011. "Pathologic lesions in chimpanzees (*Pan troglodytes schweinfurthii*) from Gombe National Park, 2004–2010." *Journal of Zoo and Wildlife Medicine,* 42: 597–607.

Terio, K. A., E. V. Lonsdorf, M. J. Kinsel, J. Raphael, I. Lipende, A. Collins, Y. Li, B. H. Hahn, D. A. Travis, and T. R. Gillespie. 2016. "Oesophagostomiasis in non-human primates of Gombe National Park, Tanzania." *American Journal of Primatology,* advance online publication, doi:10.1002/ajp.22572.

Tomasello, M. 1996. "Do apes ape?" In *Social Learning in Animals: The Roots of Culture,* edited by C. Heyes and B. Galef Jr., 319–46. New York: Academic Press.

Toth, N., and K. D. Schick. 2009. "The Oldowan: The tool making of early hominins and chimpanzees compared." *Annual Review of Anthropology,* 38: 289–305.

Toth, N., K. D. Schick, E. S. Savage-Rumbaugh, R. A. Sevcik, and D. M. Rumbaugh. 1993. "*Pan* the tool-maker: Investigations into the stone tool-making and tool-using capabilities of a bonobo (*Pan paniscus*)." *Journal of Archaeological Science,* 20: 81–91.

Townsend, S. W., T. Deschner, and K. Zuberbühler. 2008. "Female chimpanzees use copulation calls flexibly to prevent social competition." *PLoS ONE,* 3 (6): e2431.

———. 2011. "Copulation calls in female chimpanzees (*Pan troglodytes schweinfurthii*) convey identity but do not accurately reflect fertility." *International Journal of Primatology,* 32: 914–923.

Townsend, S. W., K. E. Slocombe, M. Emery Thompson, and K. Zuberbuhler. 2007. "Female-led infanticide in wild chimpanzees." *Current Biology,* 17: R355–R356.

Tyson, E. (1699) 1972. "Orang-outang sive *Homo sylvestris:* Or the anatomy of pygmie." In *Climbing Man's Family Tree,* edited by T. D. McCown and K. A. R. Kennedy, 41–48. Englewood Cliffs, N.J.: Prentice Hall.

Uehara, S. 1986. "Sex and group differences in feeding on animals by wild chimpanzees in the Mahale Mountains National Park, Tanzania." *Primates,* 27: 1–13.

Uehara, S., T. Nishida, M. Hamai, T. Hasegawa, H. Hayaki, M. Huffman, K. Kawanaka, et al. 1992. "Characteristics of predation by the chimpanzees in the Mahale Mountains National Park, Tanzania." In *Topics in Primatology,* vol. 1, *Human Origins,* edited by T. Nishida, W. C. McGrew, P. Marler, M. Pickford, and F. B. M. de Waal, 143–58. Tokyo: University of Tokyo Press.

Val, A. 2016. "Deliberate body disposal by hominins in the Dinaledi Chamber, Cradle of Humankind, South Africa?" *Journal of Human Evolution,* 96: 145–148.

Varki, A., and T. K. Atheide. 2005. "Comparing the human and chimpanzee genomes: Searching for needles in a haystack." *Genome Research,* 15: 1746–1758.

Vigilant, L., M. Hofreiter, H. Seidel, and C. Boesch. 2001. "Paternity and relatedness in wild chimpanzee communities." *Proceedings of the National Academy of Sciences,* 98: 12890–12895.

Wakefield, M. L. 2013. "Social dynamics among females and their influence on social structure in an East African chimpanzee community." *Animal Behaviour,* 85: 1303–1313.

Wallis, J. 1982. "Sexual behavior of captive chimpanzees (*Pan troglodytes*): Pregnant versus cycling females." *American Journal of Primatology*, 3: 77–88.

———. 1992. "Chimpanzee genital swelling and its role in the pattern of sociosexual behavior." *American Journal of Primatology*, 28: 101–113.

———. 1995. "Seasonal influence on reproduction in chimpanzees of Gombe National Park." *International Journal of Primatology*, 16: 435–451.

———. 1997. "A survey of reproductive parameters in the free-ranging chimpanzees of Gombe National Park." *Journal of Reproduction and Fertility*, 109: 297–307.

Washburn, S. L. 1968. "Speculation on the problem of man's coming to the ground." In *Changing Perspectives on Man*, edited by B. Rothblatt, 191–206. Chicago: University of Chicago Press.

Washburn, S. L., and C. Lancaster. 1968. "The evolution of hunting." In *Man the Hunter*, edited by R. B. Lee and I. DeVore, 293–303. Chicago: Aldine.

Watts, D. P. 1998. "Coalitionary mate guarding by male chimpanzees at Ngogo, Kibale National Park, Uganda." *Behavioral Ecology and Sociobiology*, 44: 43–55.

———. 2000. "Grooming between male chimpanzees at Ngogo, Kibale National Park: II. Influence of male rank and possible competition for partners." *International Journal of Primatology*, 21: 211–238.

———. 2004. "Intracommunity coalitionary killing of an adult male chimpanzee at Ngogo, Kibale National Park, Uganda." *International Journal of Primatology*, 25: 507–521.

———. 2007. "Effects of male group size, parity, and cycle stage on female chimpanzee copulation rates at Ngogo, Kibale National Park, Uganda." *Primates*, 48: 222–231.

———. 2008. "Scavenging by chimpanzee at Ngogo and the relevance of chimpanzee scavenging to early hominin behavioral ecology." *Journal of Human Evolution*, 54: 125–133.

———. 2010. "Dominance, power, and politics in nonhuman and human primates." In *Mind the Gap*, edited by P. M. Kappeler and J. B. Silk, 109–138. New York: Springer.

———. 2015. "Mating behavior of adolescent male chimpanzees (*Pan troglodytes*) at Ngogo, Kibale National Park, Uganda." *Primates*, 56: 163–172.

———. 2016. "Production of grooming-associated sounds by chimpanzees (*Pan troglodytes*) at Ngogo: Variation, social learning, and possible functions." *Primates*, 57: 61–72.

Watts, D. P., and S. J. Amsler. 2013. "Chimpanzee–red colobus encounter rates show a red colobus population decline associated with predation by chimpanzees at Ngogo." *American Journal of Primatology*, 75: 927–937.

Watts, D. P., and J. C. Mitani. 2000. "Infanticide and cannibalism by male chimpanzees at Ngogo, Kibale National Park, Uganda." *Primates,* 41: 357–365.

———. 2001. "Boundary patrols and intergroup encounters in wild chimpanzees." *Behaviour,* 138: 299–327.

———. 2002. "Hunting behavior of chimpanzees at Ngogo, Kibale National, Park, Uganda." *International Journal of Primatology,* 23: 1–28.

———. 2015. "Hunting and prey switching by chimpanzees (*Pan troglodytes schweinfurthii*) at Ngogo." *International Journal of Primatology,* 36: 728-748.

Watts, D. P., J. C. Mitani, and H. M. Sherrow. 2002. "New cases of intercommunity infanticide by male chimpanzees at Ngogo, Kibale National Park, Uganda." *Primates,* 43: 263–270.

Watts, D. P., M. N. Muller, S. J. Amsler, G. Mbabazi, and J. C. Mitani. 2006. "Lethal intergroup aggression by chimpanzees in Kibale National Park, Uganda." *American Journal of Primatology,* 68: 161–180.

Watts, D. P., K. B. Potts, J. S. Lwanga, and J. C. Mitani. 2012a. "Diet of chimpanzees (*Pan troglodytes schweinfurthii*) at Ngogo, Kibale National Park, Uganda, 1: Diet composition and diversity." *American Journal of Primatology,* 74: 114–129.

———. 2012b. "Diet of chimpanzees (*Pan troglodytes schweinfurthii*) at Ngogo, Kibale National Park, Uganda, 2: Temporal variation and fallback foods." *American Journal of Primatology,* 74: 130–144.

White, T. D., B. Asfaw, Y. Beyene, Y. Haile-Selassie, C. O. Lovejoy, G. Suwa, and G. Woldegabriel. 2009. "*Ardipithecus ramidus* and the paleobiology of early hominids." *Science,* 326: 75–86.

White, T. D., C. O. Lovejoy, B. Asfaw, J. P. Carlson, and G. Suwa. 2015. "Neither chimpanzee nor human, *Ardipithecus* reveals the surprising ancestry of both." *Proceedings of the National Academy of Sciences,* 112: 4877–4884.

Whiten, A. 2014. "Incipient tradition in wild chimpanzees." *Nature,* 514: 178–179.

Whiten, A., J. Goodall, W. C. McGrew, T. Nishida, V. Reynolds, Y. Sugiyama, C. E. G. Tutin, R. W. Wrangham, and C. Boesch. 1999. "Cultures in chimpanzees." *Nature,* 399: 682–685.

———. 2001. "Charting cultural variation in chimpanzees." *Behaviour,* 138: 1481–1516.

Whiten, A., K. Schick, and N. Toth. 2009. "The evolution and cultural transmission of percussive technology: Integrating evidence from palaeoanthropology and primatology." *Journal of Human Evolution,* 57: 420–435.

Wilfried, E. E. G., and J. Yamagiwa. 2014. "Use of tool sets by chimpanzees for multiple purposes in Moukalaba-Doudou National Park, Gabon." *Primates,* 55: 467–472.

Williams, J. M., E. V. Lonsdorf, M. L. Wilson, J. Schumacher-Stankey, J. Goodall, and A. E. Pusey. 2008. "Causes of death in the Kasekela chimpanzees of Gombe National Park, Tanzania." *American Journal of Primatology*, 70: 766–777.

Williams, J. M., G. Oehlert, J. Carlis, and A. Pusey. 2004. "Why do male chimpanzees defend a group range?" *Animal Behaviour*, 68: 523–532.

Wilson, A. C., and M. C. King. 1975. "Evolution at two levels in humans and chimpanzees." *Science*, 188: 107–116.

Wilson, M. L., C. Boesch, B. Fruth, T. Furuichi, I. C. Gilby, C. Hashimoto, C. L. Hobaiter, et al. 2014. "Lethal aggression in *Pan* is better explained by adaptive strategies than human impacts." *Nature*, 513: 414–419.

Wilson, M. L., M. D. Hauser, and R. W. Wrangham. 2001. "Does participation in intergroup conflict depend on numerical assessment, range location, or rank for wild chimpanzees?" *Animal Behaviour*, 61: 1203–1216.

———. 2007. "Chimpanzees (*Pan troglodytes*) modify grouping and vocal behaviour in response to location-specific risk." *Behaviour*, 144: 1621–1653.

Wilson, M. L., S. M. Kahlenberg, M. Wells, and R. W. Wrangham. 2012. "Ecological and social factors affect the occurrence and outcomes of intergroup encounters in chimpanzees." *Animal Behaviour*, 83: 277–291.

Wilson, M. L., W. R. Wallauer, and A. E. Pusey. 2004. "New cases of intergroup violence among chimpanzees in Gombe National Park, Tanzania." *International Journal of Primatology*, 25: 523–549.

Wingfield, J. C. 1984. "Androgens and mating systems: Testosterone-induced polygyny in normally monogamous birds." *Auk*, 101: 665–671.

Wittig, R. M., and C. Boesch. 2003. "Food competition and linear dominance hierarchy among female chimpanzees of the Taï National Park." *International Journal of Primatology*, 24: 847–867.

Wittig, R. M., C. Crockford, A. Weltring, T. Deschner, and K. Zuberbuehler. 2015. "Single aggressive interactions increase urinary glucocorticoid levels in wild male chimpanzees." *PLoS ONE*, 10 (2): e0118695.

Wittiger, L., and C. Boesch. 2013. "Female gregariousness in Western Chimpanzees (*Pan troglodytes verus*) is influenced by resource aggregation and the number of females in estrus." *Behavioral Ecology and Sociobiology*, 67: 1097–1111.

Wood, B. M., D. P. Watts, J. C. Mitani, and K. E. Langergraber. 2016. "Low mortality rates among Ngogo chimpanzees: Ecological influences and evolutionary implications." *American Journal of Physical Anthropology*, 159: 338 (abstract).

Worthman, C. M., and M. J. Konner. 1987. "Testosterone levels change with subsistence hunting effort in !Kung San men." *Psychoneuroendocrinology*, 12: 449–458.

Wrangham, R. W. 1977. "Feeding behaviour of chimpanzees in Gombe National Park, Tanzania." In *Primate Ecology*, edited by T. H. Clutton-Brock, 503–538. London: Academic Press.

———. 1980. "An ecological model of female-bonded primate groups." *Behaviour*, 75: 262–292.

———. 1996. *Demonic Males*. Boston: Houghton-Mifflin.

———. 1999. "Evolution of coalitionary killing." *Yearbook of Physical Anthropology*, 42: 1–30.

Wrangham, R. W., N. L. Conklin, G. Etot, J. Obua, K. D. Hunt, M. D. Hauser, and A. P. Clark. 1993. "The value of figs to chimpanzees." *International Journal of Primatology*, 14: 243–256.

Wrangham, R. W., N. L. Conklin-Brittain, and K. D. Hunt. 1998. "Dietary response of chimpanzees and cercopithecines to seasonal variation in fruit abundance: I. Antifeedants." *International Journal of Primatology*, 19: 949–970.

Wrangham, R. W., K. Koops, Z. P. Machanda, S. Worthington, A. B. Bernard, N. F. Brazeau, R. Donovan, et al. 2016. "Distribution of a chimpanzee social custom is explained by matrilineal relationship rather than conformity." *Current Biology*, 26: 1–5.

Wrangham, R. W., and E. van Zinnicq Bergmann-Riss. 1990. "Rates of predation on mammals by Gombe chimpanzees, 1972–1975." *Primates*, 31: 157–170.

Wrangham, R. W., and M. L. Wilson. 2004. "Collective violence: Comparisons between youths and chimpanzees." *Annals of the New York Academy of Sciences,* 1036: 233–256.

Wrangham, R. W., M. L. Wilson, and M. N. Muller. 2006. "Comparative rates of violence in chimpanzees and humans." *Primates,* 47: 14–26.

Wroblewski, E. E., C. M. Murray, B. F. Keele, J. C. Schumacher-Stankey, B. H. Hahn, and A. E. Pusey. 2009. "Male dominance rank and reproductive success in chimpanzees, *Pan troglodytes schweinfurthii*." *Animal Behaviour*, 77: 873–885.

Yamakoshi, G., and Y. Sugiyama. 1995. "Pestle-pounding behavior of wild chimpanzees at Bossou, Guinea: A newly observed tool-using behavior." *Primates*, 36: 489–500.

Zamma, K. 2002. "Leaf-grooming by a wild chimpanzee in Mahale." *Primates*, 43: 87–90.

———. 2013. "What makes wild chimpanzees wake up at night?" *Primates*, 55: 51–57.

Acknowledgments

I began this book much the way I've written each of my books: as a learning experience for myself. Several years after the conclusion of my last field research on wild chimpanzees, I felt I was falling behind on recent developments. I had moved on to new projects on other animal species, and chimpanzee research wasn't on my mind as it had long been. In an earlier time, updating myself would have required some extended reading in the periodical room of my university library, followed by hours spent standing in front of a balky machine making photocopies of the articles for my files. Now it meant reading the past decade's worth of digital publishing about chimpanzees from the comfort of my desk. As I read papers—some by old colleagues and friends and others by a new generation of young scientists—what impressed me was how far we have come. First, there are many more field studies. My colleague William McGrew recently tallied 120 chimpanzee field studies that have produced at least one scientific publication. Most of these were short term, but nearly a dozen studies have been conducted over decades and are ongoing. There is far more detailed information on everything from vocalizations to ovulation than ever before. Laboratory analyses of sex hormones, DNA, and vocal acoustics now augment time-honored observation methods. We can estimate diets of unseen apes using radioisotope analysis of their shed hairs. We can extract information on reproductive state, energy deficit, and paternity from their feces or urine. New theories have sought to explain everything from intercommunity violence to baby rearing. So I sat down to write it all up in a form that would be interesting and useful both to the community of primatologists who study great apes and to an enlightened readership at large. As I wrote the book, more exciting research findings emerged. So I chose a time frame and synthesized all that has happened within it: chimpanzee field research since the late 1990s, or roughly the past twenty years.

I dedicate *The New Chimpanzee* to Jane Goodall, and to the generations of primatologists who have followed in her footsteps. Many of us in the chimpanzee research community joke about our relationship to Jane using a pun on the title of her most acclaimed early book, *In the Shadow of Man*. We all live and work in the shadow of Jane. Her pioneering research and enduring dedication, to both primate conservation and the environmental crises facing our planet, are sources of inspiration to all of us. Her early work influenced our view of great apes more than that of all other primatologists combined. Goodall's project in Gombe National Park is now approaching its sixtieth year, the longest and arguably most important study of any primate or other wild animal ever conducted. Of course, Goodall didn't carry the project alone through all those years. Before, during, and following my time at Gombe, Anthony Collins helped to guide Gombe research and conservation. More recently, so have Shadrack Kamenya, Michael Wilson, and others.

My research at Gombe in the 1990s would not have been possible without the hard work of Tanzanian research assistants. They included, during my time at Gombe, Karoli Alberto, Yahaya Almasi, Bruno Herman, Madua Juma, the late Msafiri Katoto, Hilali Matama, Tofficki Mikidaddi, Hamisi Mkono, Eslom Mpongo, David Mussa, Gabo Paulo, Nasibu Sadiki, Issa Salala, Methodi Vyampi, and Selemani Yahaya. I am also grateful to the offices of Tanzania National Parks and the Tanzanian Commission for Science and Technology for granting research permissions during those years, and to various granting bodies, including the National Science Foundation, the National Geographic Society, and the Leakey Foundation, for their support.

During my years studying the coexistence of chimpanzees and mountain gorillas in Bwindi Impenetrable National Park, Uganda, I relied on the help of many Ugandans and expatriates. Dr. John Bosco Nkurunungi, my former PhD advisee, was a research assistant, then a graduate student, and later a colleague and university lecturer, as well as a friend. Gervase Tumwebaze was hired as an inexperienced field assistant. He grew to be the local leader on the project and eventually a politician and leader in the local Nkuringo community. For many years of permission to conduct research in Bwindi, I thank the Uganda Wildlife Authority, the Ugandan National Council for Science and Technology, and the Institute of Tropical Forest Conservation. I am grateful that during those years I had the support of Drs. Michele Goldsmith, Richard Malenky, Alastair McNeilage, Martha Robbins, and Nancy Thompson-Handler. I also thank Keith Masana and Bwindi warden Christopher Oreyema, as well as Ambrose Ahimbisibwe; Robert Berygera; Eric Edroma; Fenni Gongo and his father, Mzee Gongo; Simon Jennings; Johanna Maughn; Evarist Mbonigaba; Caleb Mgambaneza; and Senior Ranger Silva Tumwebaze, among others. I owe special thanks to Mitchell Keiver, who spent a year making the project run, punctuated by the hardship he endured during the rebel attack in 1999.

The global community of chimpanzee field researchers is fairly large these days, but we stand on the shoulders of a handful of senior scientists whose work inspired and informed us. They may not be household names like Goodall, but their long-term efforts to collect basic information, to generate theories to explain what they have observed, and to guide the careers of younger colleagues have made them central to my career and that of many others. They're the researchers whose work set the standard for all others. I have cited each of these individuals many times in the book. I've also argued with some of them, both in print and in person (politely, usually), because that is the nature of doing science. For advice and critiques over the years, I thank Christophe Boesch, William McGrew, the late Toshisada Nishida, Anne Pusey, and Richard Wrangham.

As always, this book is also the collaborative effort of many people whose names will appear only in these acknowledgments. Some have read chapters of the manuscript, and others have served as sounding boards over the years for discussing the issues in it. I thank my colleagues John Allen, Christopher Boehm, Stephanie Bogart, and Joe Hacia. I've also been very fortunate to have mentored a succession of wonderful doctoral students who are now my colleagues: James Askew, Andrea Currylow, Andrew Fogel, Angela Garbin, Jess Hartel, R. Adriana Hernandez, Xuecong Liu, Laura Loyola, Maureen McCarthy, Martin Muller, John Bosco Nkurunungi, Robert O'Malley, Norm Rosen, Michael Tuma, and Pratyaporn Wanchai.

This is my fifth book with Harvard University Press, and I am indebted to my former editor Michael Fisher and current editor Andrew Kinney for their support of the manuscript. I thank Olivia Woods, Mary Ribesky, and Stephanie Vyce for their help. Multiple anonymous reviewers critiqued the manuscript in various stages of its preparation and corrected my innumerable errors and awkward passages. I thank Marika Stanford-Moore for help in preparing the figures.

My children have grown up tolerating my long absences from home and my daily long hours locked in my home office writing. They've reached the age at which they can be among my best critics. So I thank Gaelen, Marika, and Adam. And, as always, I thank my wife, Erin, for a lifetime of love and support.

Credits

Page

13 Base map from http://d-maps.com/carte.php?num_car=736&lang=en. Copyright © d-maps.com

50 From Wroblewski et al. 2009, figure 1. Copyright © 2009, The Association for the Study of Animal Behaviour. Published by Elsevier Ltd. All rights reserved. Reprinted with permission from Elsevier.

51 From Wroblewski et al. 2009, figure 3. Copyright © 2009, The Association for the Study of Animal Behaviour. Published by Elsevier Ltd. All rights reserved. Reprinted with permission from Elsevier.

61 From Pusey, Williams, and Goodall 1997, figure 3. Copyright © 1997, The American Association for the Advancement of Science. Reprinted with permission from AAAS.

63 From Murray, Mane, and Pusey 2007, figure 3. Copyright © 2007, The Association for the Study of Animal Behaviour. Published by Elsevier Ltd. All rights reserved. Reprinted with permission from Elsevier.

93 Redrawn from Emery Thompson 2013, figure 1.

126 From Muller and Wrangham 2014, figure 2. Copyright © 2013, Elsevier Ltd. All rights reserved. Reprinted with permission from Elsevier.

128 From Muller and Wrangham 2014, figure 3. Copyright © 2013, Elsevier Ltd. All rights reserved. Reprinted with permission from Elsevier.

169 From Pruetz et al. 2015, figure 2. Copyright © 2015, The Authors (CC-BY-4.0).

Index

Acclimation, 4–5
Adaptive violence, 67–69
Adolescence, 123–125
Adoption, 111, 121–123
Aggression. *See* Violence and aggression
Aging, 110, 125–129
Ahlquist, Jon, 178
Akeley, Carl, 8
Alarm calls, 38
Alberts, Susan, 48
Alpha status, 43–44, 48–49, 51–54, 63–65, 94, 98–99
Altruistic behavior, 121–122
Anthropomorphism, 7
Ants, 166–168
Ape meat, 202
Apolipoprotein E (*APOE*), 182–183
Arcadi, Adam Clark, 37
Archaeology, 160–162, 177
Ardipithecus ramidus, 194–196
Army ants, 166–167
Arnold, Kate, 45–46, 89
Arroyo, Adrián, 159
Atheide, Tasha, 181
Atsalis, Sylvia, 110
Australopithecus afarensis, 193, 194, 195

Baboons: diet of, 26; dominance dynamic in, 42, 46–48; hormonal bases of social behavior among, 56–57, 58; mating preferences of, 103; as prey, 134

Baculum, 95
Ban, Simone, 27
Basabose, Augustin, 16
Bates, Lucy, 76
Beethoven, 106
Behavior: interpreting, from fossils, 194; and adaptation, 200–201. *See also* Culture and cultural behaviors
Berger, Lee, 196
Bergmann-Riss, Emily van Zinnicq, 137
Bertolani, Paco, 200
Birth clustering and staggering, 101
Blackwell, Lucinda, 200
Blumenbach, Friedrich, 176
Body hair loss in humans, 182
Body size, 118
Boehm, Christopher, 66
Boesch, Christophe: research of, in West Africa, 14–15; on sociability in females, 31; on dominance and mating preference, 48; on dominance and paternity, 51–52; on female dominance, 60; on sexual coercion, 72, 105; on intercommunity relations in West Africa, 77; on attack strategies in Taï National Park, 81; on reconciliation, 89; on bystanders, 90; on female migration, 109; on maternal investment in offspring, 117; on adoption of infants, 121–122; on hunting, 135, 141, 150; on meat sharing and mating, 151; on tool use, 158, 159, 163

Index

Boesch-Achermann, Hedwige, 159
Bogart, Stephanie, 164
Bonobos: diet of, 24; female alliances in, 73; tool use in, 157, 161, 199; and distinctiveness of chimpanzees as species, 177; genetic divergence of chimpanzees and, 179, 187
Border patrols, 75–77
Bossou, Republic of Guinea, 14, 170–171
Botero, Maria, 121
Budongo Forest Reserve: study site in, 13–14; alarm calls in, 38; grooming in, 45–46; dominance in, 54–55, 60; intracommunity violence in, 70, 73; adoption of orphans in, 122; social transmission in, 156; tool use in, 173–174. *See also* Sonso community
Bulindi chimpanzees, 163
Burial, 197–198
Bushmeat crisis, 202
Busse, Curt, 134
Bustamante, Carlos, 181
B-vitamin, 146
Bwindi Impenetrable National Park: study of chimpanzees in, 16; social interactions in, 19–21; nests in, 34; sleeping and waking cycles in, 35; dominance in, 49; tool use in, 163; and conservation, 204
Bygott, David, 46
Byrne, Richard, 76
Bystanders, 90

Calories, as reason for hunt, 143–144
Captive chimpanzees: cognitive abilities of, 8, 161; reconciliation in, 88–89; swellings in, 92–94; female reproductive aging in, 110; impact of maternal deprivation on, 121; lifestyle and lifespan of, 125; tool use in, 157, 159; handedness in, 170
Charlie, 79
Chen, Feng-Chi, 179–180
Chimpanzee history, 1
Chimpanzee politics. *See* Chimpanzee society; Dominance

Chimpanzee research, 1–2, 16–18; and observing wild chimpanzees, 2–8; history of, 8; Goodall and, 9–11; in Mahale Mountains, 11–14; field studies of more than fifteen years, 12*t*; seven longest chimpanzee studies, 13*f*; in western Africa, 14–16; human sexuality and, 112–113; and protection of chimpanzees, 202–206
Chimpanzees: humans' relationship with, 1–2; personas of, 6–7, 152; classification of, 176–177; defining, as species, 184–190; discoveries regarding, 198–201; protection of, 202–206
Chimpanzee society: structure of, 12, 21–23, 29–30, 36, 39–40, 48, 58; discoveries regarding, 21–23; impact of diet on, 24–25; diet and mating in, 28–32; and presence of estrous females, 32–33; nests and understanding, 33–36; vocalizations and, 36–39; hunting and meat eating in, 131, 134, 147; meat sharing and, 150–153; genetics and understanding, 187–190. *See also* Culture and cultural behaviors; Dominance; Violence and aggression
Civets, 132
Coalitionary aggression, 29, 30, 73, 74, 77, 104
Cognition: in captive chimpanzees, 8, 161; and social structure and individual behavior, 40; in infants, 160
Colobus monkeys, 130–131, 134, 135–137, 139–141, 146, 148–149, 152
Conklin-Brittain, Nancy, 145
Conservation, 202–206
Consortships, 101–103
Cooperation: among male chimpanzees, 29–30, 188; in hunting, 149–150
Copulation calls, 96–97
Cortisols, 47, 58–59

Crockford, Catherine, 38
Culture and cultural behaviors, 154–157; diversity in, 109, 131–132, 155–156, 199; meat sharing as, 147–148; tool use as evidence of, 157–160; and chimpanzee archaeology, 160–162; and handedness, 169–172; and genetics, 172–173; and ecology, 173–175; discoveries regarding, 200–201. *See also* Chimpanzee society

Darwin, Charles, 177, 191
David Greybeard, 4
Death, 197–198
D'Errico, Francesco, 200
Deschner, Tobias, 94
De Waal, Frans, 41, 88
Diet: implications of diverse, 22; and daily foraging, 24–28; and mating, 28–32, 39; correlation among party size and estrous females and, 33, 39; and migration, 109; meat in, 132–133, 136–137; of early humans, 200. *See also* Food; Foraging; Hunt and hunting
Dinaledi Chamber, 196–198
Dirks, Paul, 197
Disease, 127
Dixson, Alan, 56, 95
DNA, shared, 179–184
DNA hybridization, 178
Dominance, 41–43; of males over females, 42, 55; and alpha status, 43–44; and grooming, 44–46; and mating success, 46–54; and signaling submission, 54–55; and testosterone, 55–58; and stress, 58–59; among females, 59–62; origins of, 63–65; and intracommunity violence, 70; and female mate choice, 98–99, 100; and male mating strategies, 102; male mate choice and female, 104; and female reproductive rate, 110; in reproductive skew theory, 112; and meat eating, 141
Dorus, Steven, 181

Dorylus, 166–167
Drumming, 37
Duffy, Kimberly, 101

Ebola, 127
Egg depletion, 110
Ely, John, 114
Emery Thompson, Melissa, 94, 96, 103, 106, 109–110, 111, 118
Emulation, 161
Endangered species, 205
Essentialist view of species, 176
Evolution. *See* Human-ape divergence; Human origins

Fahy, Geraldine, 140
Fallon, Brittany, 97
Fat: in ape milk, 115–116; as reason for hunt, 145–146; in insects, 166
Fedurek, Pawel, 36, 38–39, 56
Feldblum, Joseph, 44, 72, 105
Female chimpanzees: and dominance, 21, 42, 59–62; sociability of, 22, 39–40; and diet's role in mating, 28–32, 39; and party size, 32–33; and intracommunity violence, 71–73; and intercommunity violence, 77–78, 80, 81–82; and infanticide, 86–87; reproductive life of, 91, 109–112, 125; swellings in, 92–97, 108–109, 149, 151; and selectivity in mating, 97–101; and male mate choice, 103–105; migration of, 105–109, 168, 172–173, 174; milk produced by, 115–116; and maternal investment in offspring, 116–117; infant development in, 118–119; adolescent, 124; age-specific survival probability in, 126f; hunting in, 140–141; and hunting party size, 149; meat sharing and, 150–152; tool use in, 165
Fertility, 91, 94–95, 101, 109–112
Fifi, 86–87, 107–108
Fig trees, 19–21
Finch, Caleb, 52, 53, 182–183

Fischer, Anne, 186, 187
Fission-fusion social system, 21–23, 29, 36, 39–40, 48, 58
Flame, 120
Flint, 120–121
Flo, 107, 120–121
Flossi, 168
Fluid dipping, 162–164, 173
Foerster, Steffen, 59–60
Fongoli, Senegal, 17, 200–201
Food: party size and availability of, 29, 83; and female dominance, 60–61, 62; and intercommunity violence, 83, 84; tool use and, 138, 157–160, 162–169, 173–174, 200. *See also* Diet; Foraging; Fruit; Hunt and hunting
Food calls, 37–38
Foraging: daily, 23–28; territorial defense in, 76; hunting and, 141; tool use and, 157–159, 162–169, 173–174. *See also* Diet; Food; Fruit
Fossey, Dian, 203
Fossils, 7
FOXP2, 181–182
Frodo, 6, 37, 147–148
Fröhlich, Marlen, 118
Fruit: in chimpanzee diet, 23–28; correlation among party size and estrous females and availability of, 33, 39. *See also* Foraging
Fünfstück, Thomas, 186–187

Gagneux, Pascal, 179
Garner, Karen, 186
Gashaka Gumti National Park, 17
Genetics, 176–177; dating human-ape divergence, 177–179; shared DNA sequences between apes and humans, 179–184; defining chimpanzees as species, 184–190
Ghiglieri, Michael, 15
Gilby, Ian, 139, 144, 147
Gillespie, Thomas, 127
Glycans, 181
Goblin, 6, 44, 49, 64, 70
Goldberg, Anthony, 188

Gombe Stream Reserve: field studies at, 2–3, 9–11; discoveries regarding chimpanzee society at, 21–22; sociability of chimpanzees at, 31–32; dominance at, 49–51, 64; paternity and dominance at, 49–51; female dominance rank at, 59–60, 61–62; intracommunity violence at, 70, 72; intercommunity violence at, 75, 78, 84; community split in, 78–82; infanticide at, 86–87; female mate choice at, 100; male mating strategies at, 102; sexual coercion at, 105; female migration at, 107–108; reproductive output at, 109–110; maternal investment in offspring at, 117; infant development at, 118–119; social play at, 119–120; orphaned infants at, 120–121; community size of, 128; hunting and meat eating at, 133–134, 137, 139, 145, 147–148; scavenging in, 142; factors influencing party size at, 149; tool use at, 164–166; cultural transmission at, 168. *See also* Kahama community; Kasekela community
Gomes, Christina, 151
Gonder, Katherine, 186
Goodall, Jane: as beginning of modern chimpanzee field studies, 4–5; and perspectives on chimpanzee personas, 6–7; impact of, 9–11; discoveries made by, 21–22; on community split and warfare in Gombe, 75; and division of Kasekela community, 78–79; on mating, 91; on aggression in consortship, 102; observes hunting and meat eating, 131–132; on tool use, 164; and study of human origins, 191
Goodman, Morris, 179
Gorillas, 8, 9, 16, 34, 96, 116, 177, 178–179, 185–186
Goualougo Triangle, 16–17
Grandmother hypothesis, 111–112
Gremlin, 87

Grooming: dominance rank and, 42, 44–46; of leaves, 156
Group mating, 101–103
Gruber, Thibaud, 157–160, 164, 173–174

Habituation, 4–5, 31, 168
Hahn, Beatrice, 183
Handedness, 169–172
Harcourt, Alexander, 96
Hartel, Jess, 89
Hashimoto, Chie, 27–28
Haslam, Michael, 199–200
Hausfater, Glenn, 48
Hawkes, Kristen, 110, 111
Heintz, Matthew, 119–120
Hernandez-Aguilar, R. Adriana, 16, 34, 162
Hicks, Thurston, 163
Hinde, Katie, 115, 116
HIV (human immunodeficiency virus), 183–184
Hobaiter, Catherine, 122, 156
Homo habilis, 197
Homo naledi, 196–198
Honey tools, 162–164, 173
Hopkins, William, 170
Hormones, 55–59, 77
Hrdy, Sarah, 85, 93
Hugh, 79
Human-ape divergence: dating, 177–179, 189; and overlap in human and chimpanzee DNA sequence, 179–184
Human fertility, 110–111
Human immunodeficiency virus (HIV), 183–184
Human origins: study of chimpanzees and, 191–192; chimpanzees and earliest humans, 193–196; *Homo naledi*, 196–198
Humans: sexuality in, 112–113; survival probability of, 128*f*, 129; tools of early, 199–200; diet of early, 200; behavior of early, 201
Humle, Tatyana, 170–171
Humphrey, 79

Hunt, Kevin, 16, 35
Hunt and hunting: experience of, 130–131; field observations and studies on, 131–137; as seasonal, 137, 139; planning and tools in, 137–138, 169; efficiency and success in, 138–139, 140, 148–150; bursts in, 139; in males versus females, 140–141; intention in, 141–142; reasons for, 143–147, 150; factors influencing, 147–153; and understanding early humans, 195
Huxley, Thomas, 176–177

Illness, 127
Imanishi, Kinji, 9
Imitation, 161
Infants: killing of, 85–87, 101; orphaned, 111, 120–123; and grandmother hypothesis, 111–112; birth and first years of, 114–115; and nutritional value of mothers' milk, 115–116; maternal investment in, 116–117; paternal involvement with, 117–118; development of, 118–119; play in, 119–120; tool use in, 165
Inoue, Eiji, 188
Inoue-Nakamura, Noriko, 160
Insects, 162–163, 164–168, 200
Intercommunity aggression, 73–78; and Gombe community split, 78–82; explanations for, 82–85
Intervention, 6, 49, 70
Intracommunity aggression, 69–73
Itani, Junichiro, 9, 11

Jablonski, Nina, 177
Janmaat, Karline, 24
Japan, primatology in, 11
Johanson, Donald, 193
Jones, James, 110
Jungers, William, 193
Juveniles, 123–125

Kahama community, 78–82, 128
Kahlenberg, Sonya, 107

Kalan, Amie, 38
Kalcher-Sommersguter, Elfriede, 121
Kamemanfu, 44
Kamilar, Jason, 174
Kanyawara chimpanzees: diet of, 26, 27, 145; and intercommunity violence, 81, 83–84; male mate choice in, 103; sexual coercion among, 104–105; female migration among, 108–109; juveniles among, 123; life expectancy of, 126; hunting and, 144; tool use in, 173–174; relatedness among, 188
Kanzi, 161, 199
Kasekela community, 78–82, 168
Keele, Brandon, 183
Ketones and Ketosis, 144
Kibale Chimpanzee Project, 15
Kibale National Park, 26, 81
Kidevu, 20
Killings. *See* Violence and aggression
King, Mary Claire, 181, 184
Kinship: and male cooperation, 29, 188; and social bonds, 45, 187–188; and political relationship among males, 47; and meat sharing, 134, 150; immunological, 177
Koops, Kathelijne, 34, 167
Kortlandt, Adrian, 9
Kret, Mariska, 92
Kroeber, Alfred, 154–155
Kühl, Hjalmar, 171–172

Lactation, 115–116
Lancaster, Chet, 151–152
Langergraber, Kevin, 39, 106, 112–113, 172–173, 174–175, 184, 189
Language, 181–182
Langur infanticide, 85–86
Laporte, Marion, 55
Leakey, Louis, 9, 10, 191
Leaves, 23; clipping, 156; grooming, 156
Lehmann, Julia, 31, 117
Li, Wen-Hsiung, 179–180
Lice, 182
Life expectancy, 52–54, 110–111, 125–129, 183

Linnaeus, Carolus, 176
Loango National Park, 163–164
Lonsdorf, Elizabeth, 118–119, 164–165, 170
Lovejoy, C. Owen, 193, 194–196
Lucy, 193
Luit, 41
Luncz, Lydia, 109
Lycett, Stephen, 172

Machanda, Zarin, 45
Mahale Mountains National Park: field studies in, 11–14; discoveries regarding chimpanzee society at, 22; party size in, 32–33; sleeping and waking cycles in, 35; grooming in, 46; intracommunity violence in, 70; intercommunity violence in, 74–75, 81–82; mating in, 98–99, 100; birth staggering at, 101; disease outbreak in, 127; hunting in, 135; hunting and meat eating in, 137; relatedness among male chimpanzees at, 188
Male chimpanzees: aggression in, 6, 57, 66, 67–68; and role of dominance in mating, 21; cooperation among, 29–30, 188; social behavior of, 39; grooming in, 45; origins of dominance in, 63–64; and intracommunity violence, 69–71; and intercommunity violence, 73, 75–76, 77, 79–82, 84–85; and infanticide, 86; genitalia size of, 95–96; and female selectivity in mating, 97–101; mating strategies of, 101–103, 150–152; mate choice in, 103–105; migration of, 106; acceptance of immigrant females, 107; bonds with females, 112–113; and maternal investment in offspring, 116–117; and infant care, 117–118; infant development in, 118–119; adoption of orphans by, 122; adolescent, 124–125; age-specific survival probability in, 126f; hunting and, 140–141, 148–149; meat sharing and, 150–152; tool use in, 165;

relatedness among, 187–189.
  See also Dominance
Man the Hunter theory, 151–152
Marshack, Joshua, 68, 174
Martha, 20
Mating and reproduction: and dominance, 21, 42, 47–52, 61–62, 64; and diet, 28–32, 39; and intracommunity violence, 69, 71–72; and female life cycle, 91; and female swellings, 92–97; choosiness and choice in, 97–101, 103–105; group versus consort, 101–103; and female migration, 105–109; and female fertility decline, 109–112; human sexuality and chimpanzee research, 112–113; and hunting and meat sharing, 149, 150–151; monogamy in, 196
Matsumoto-Oda, Akiko, 32–33, 98–99, 101
Matsuzawa, Tetsuro, 160, 170–171
Mboneire, 20, 49
McBrearty, Sally, 177
McCarthy, Maureen, 25, 53, 190
McGrew, William, 15, 154–155, 166, 200
McLennan, Matthew, 163
Meat eating. *See* Hunt and hunting
Menopause, 111–112
Menstrual cycle, 92, 93f, 94
Mercader, Julio, 161
Migration, 105–109, 168, 172–173, 174
Milk, 115–116
Milligan, Laura, 115, 116
Mitani, John: restarts Ngogo project, 15; on pant hoot calls, 37; on chimpanzee social network, 45; on intercommunity aggression, 76, 77, 85; on hunting, 140, 144, 149; on meat sharing, 147
Mitumba community, 11, 80, 168
Molecular genetics, 177–178, 190
Monogamy, 196
Moore, Jim, 16
Moorjani, Priya, 184

Morality, of aggression, 67
Morgan, David, 16–17
Mormon settlers, fertility curves for, 110–111
Mortality, 52–54, 125–129
Moss-sponges, 156–157
Muehlenbein, Michael, 47
Muller, Martin: on cortisol and dominance rank, 47; on aggression and hormones, 57, 58–59; on sexual coercion, 71, 104–105; on intercommunity territoriality, 84; on male mate choice, 103; on life expectancy and mortality, 126, 128; on hunting, 144
Murder, 69. *See also* Violence and aggression
Murray, Carson, 59, 86
Musgrave, Stephanie, 160
Mutual grooming, 45

Nakamura, Michio, 121
Nests, 33–36
Newton-Fisher, Nicholas, 27, 54–55, 73
Ngogo chimpanzee community: study of, 15; diet of, 25, 26; female gregariousness in, 32; grooming in, 46; intracommunity violence in, 70–71; border controls at, 76; and intercommunity violence, 77, 81, 84; mating in, 99; female mate choice at, 100; mate guarding in, 104; life expectancy in, 126–127; hunting and meat eating in, 140, 147; scavenging in, 142; tool use in, 173–174
N'guessan, Antoine, 27
Nishida, Toshisada, 11–12, 22, 43–44, 78, 100, 121, 135
Ntologi, 43–44, 188
Nut cracking, 159–160
Nutrition: of ape milk, 115–116; as reason for hunt, 146–147

Oil palm trees, 145–146
Oken, Lorenz, 176
Olduvai Gorge, 139

Olshansky, S. Jay, 52–53
O'Malley, Robert, 151, 165–166, 168
Orphans, 111, 120–123
Ovulation, 92, 93–94, 95, 99, 100–101

Palm trees and fruit, 145–146
Pant grunts, 54–55, 60
Pant hoots, 36–37
Pan troglodytes, 176–177, 186
Party size: fruit abundance and, 27–28; and food availability, 29, 83; and sociability, 31; and presence of estrous females, 32–33, 39; as factor in decision to hunt, 139; and hunting success, 148–149
Passion, 87
Paternity: and dominance, 21, 47–48, 49–52, 98, 188; confusing, 72, 93; and female mate choice, 97–101; and aggression, 114
Patterson, Nick, 189
Peacemaking, 87–90
Pecking order, 42
Penis size, 95
Personas, of chimpanzees, 6–7, 152
Plasticity, biological, 194
Play, in infants, 119–120
Plooij, Frans, 118
Pom, 87
Pontzer, Herman, 123
Potts, Kevin, 26
Power, Margaret, 68
Power, Michael, 165–166
Pradhan, Gauri, 82
Prado-Martinez, Javier, 185
Pregnancy, female sexual behavior during, 93
Presidents, life spans of U.S., 52–53
Prof, 131
Protein, 144–145, 166, 200
Pruetz, Jill, 17, 138, 164, 169, 200
Prüfer, Kay, 185
Przeworski, Molly, 184
P. t. ellioti, 186
P. t. schweinfurthii, 186
P. t. troglodytes, 186

P. t. verus, 186
Pusey, Anne, 59–60, 61–62

Reconciliation, 87–90
Red colobus monkeys, 130–131, 134, 135–137, 140, 148, 152
Reproduction. *See* Mating and reproduction
Reproductive skew theory, 112
Research. *See* Chimpanzee research
Reynolds, Vernon, 13
Rogers, Alan, 182
Rumbaugh, Duane, 161
Ruvolo, Mary Ellen, 179
Ryder, Oliver, 186

Salt, 146
Samson, David, 35
Samuni, Liran, 74
Sandel, Aaron, 42
Sanz, Crickette, 16–17, 162
Sapolsky, Robert, 47, 56
Sarich, Vincent, 177, 178
Sarmiento, Esteban, 185–186
Savage-Rumbaugh, Sue, 161
Sayers, Ken, 194, 195
Scavenging, 137–138, 142
Schaller, George, 9
Schick, Kathy, 161, 199
Schöning, Caspar, 167
Schrauf, Cornelia, 160
Sex ratio evolution, 116–117
Sexual coercion, 71–72, 104–105
Sexuality, human, 112–113
Sherrow, Hogan, 43
Sibley, Charles, 178
Simian immunodeficiency virus (SIV), 127–128, 183–184
Sirianni, Giulia, 159
SIV (simian immunodeficiency virus), 127–128, 183–184
Skin pigment, 182
Sleep, 33–36
Smith, Ken, 110
Smuts, Barbara, 48
Sobolewski, Marissa, 77

Social harmony, 87–89
Social learning, 154, 155, 160, 164–165. *See also* Culture and cultural behaviors
Society. *See* Chimpanzee society
Sommer, Volker, 17
Sonso community: fruit abundance and distribution in, 27; dominance in, 60; female coalitions in, 73; reconciliation in, 89; and female migration, 106; violence against immigrant females in, 107; adoption patterns in, 122–123; social transmission in, 156–157; tool use in, 174
Species: classification of, 176–177; defining chimpanzees as, 184–190; endangered, 205. *See also* Human-ape divergence
Sperm competition, 96
Stanford, Craig: on amount of meat eaten, 137; on factors influencing, 147–153; on *Homo naledi*, 195–198; on hunting efficiency, 138; on hunting frequency, 139; on scavenging, 142; starts hunting research, 135
Stern, Jack, 193
Stewart, Fiona, 36
Stone throwing, 171–172, 201
Stress, 58–59
Strum, Shirley, 48
Stumpf, Rebecca, 48, 72, 99–100, 108
Style, 155
Submission, 54–55
Subspecies, 185–187
Sugiyama, Yukimaru, 14
Susman, Randall, 193
Sussman, Robert, 68
Swellings, 92–97, 108–109, 149, 151. *See also* Mating and reproduction

Taï National Park: field studies in, 14–15; foraging in, 24; chimpanzee diet in, 27; sociability of chimpanzees in, 31, 32; nests in, 34; food calls in, 38; dominance and paternity in, 51–52; intracommunity violence in, 72; intercommunity violence in, 77–78, 81; reconciliation in, 89; female mate choice in, 99–100; sexual coercion in, 105; maternal investment in offspring in, 117; paternal involvement with infants in, 117; adoption of orphans in, 121–122; disease outbreak in, 127; hunting and meat eating in, 135, 136, 137, 145, 150; tool use in, 158–159, 161–162; relatedness among male chimpanzees in, 188
Takahata, Yukio, 75
Takemoto, Hiroyuki, 187
Teleki, Geza, 132–134, 147
Tennie, Claudio, 143
Terio, Karen, 127
Termite fishing, 164–165, 200
Testes size, 95–96
Testosterone, 55–58, 77
Throwing, 171–172, 201
Tomasello, Michael, 161
Tomonaga, Masaki, 92
Tools: and cultural diversity, 109; and food acquisition, 138, 157–160, 162–169, 173–174, 200; as evidence of chimpanzee culture, 156–160; and chimpanzee archaeology, 160–162; and handedness, 169–172; and genetics, 172–173; and ecology, 173–175; of hominins, 199–200
Toro-Semliki Wildlife Reserve, 16, 34, 35
Toth, Nicholas, 161, 199
Townsend, Simon, 96–97
Trace minerals and elements, 146–147
Traditions, 155–156
Trezia, 168
Tubers, 162
Twins, 114
Tyson, Edward, 8, 176

Ugalla chimpanzees, 162
U.S. presidents, life spans of, 52–53
Uvinza, Tanzania, 16

Val, Aurore, 198
Van Roosmalen, Marc, 88
Varki, Ajit, 181
Videan, Elaine, 110
Vigilant, Linda, 172–173, 188
Violence and aggression: in male chimpanzees, 6, 57; stress hormones and, 58–59; use of, 66; origins of, 67–69; intracommunity, 69–73; sexual, 71–72, 104–105; intercommunity, 73–78; and war between Kasekela and Kahama communities, 78–82; explanations for, 82–85; against infants, 85–87; reconciliation following, 87–90; in consortships, 102; against immigrant females, 107
Vocalizations, 36–39, 54–55, 60, 96–97, 118

Wakefield, Monica, 32
Wallis, Janette, 92–93, 100
Washburn, Sherwood, 9, 151–152, 191–192
Watts, David: restarts Ngogo project, 15; on foraging and diet, 25, 26; on grooming and dominance rank, 46; on cortisol and dominance rank, 47; on intracommunity violence, 70, 71; on intercommunity aggression and territoriality, 76, 77, 84; on average mating rate of females, 92; on female sexual behavior, 99; on female mate choice, 100; on mate guarding, 104; on hunting, 140, 144, 149; on scavenging, 142; on meat sharing, 147

Weapons, 66, 138, 169
White, Tim, 193
Whiten, Andrew, 45–46, 89, 155–156, 199
Whitten, Patricia, 94
Williams, Jennifer, 82, 84, 85
Wilson, Allan, 177
Wilson, Michael, 68, 83–84, 86, 178, 181, 184
Wingfield, John, 56
Wittig, Roman, 58, 60, 89, 90
Wood, Brian, 126–127
Wrangham, Richard: and Kibale Chimpanzee Project, 15; on diet, 22, 27, 29, 145; on mating, 29; on party size and fruit availability, 29; on cortisol and dominance rank, 47; on aggression and hormones, 57, 58–59; on violence among wild chimpanzees, 66; on community split and warfare in Gombe, 75; on intercommunity attacks, 82–83; on infanticide, 86; on male mate choice, 103; on juveniles, 123; on life expectancy and mortality, 126, 128; on meat consumption, 137; on hunting, 144; on style, 155; on relatedness among males, 188
Wroblewski, Emily, 49, 50

Yamagiwa, Juichi, 16
Yerkes, Robert, 191
Yeroen, 41

Zamma, Koichiro, 35
Zuberbühler, Klaus, 13–14, 55